DISCOURSE ON

FLOATING BODIES

TO THE MOST SERENE
Don Cosimo II.
GREAT DUKE *OF* TUSCANY,
CONCERNING
The *NATATION* of BODIES Vpon,
And *SUBMERSION* In THE WATER.

by
Galileus Galilei

Philosopher and Mathematician
unto His most Serene Highnesse.

Printed on acid free ANSI archival quality paper.
ISBN: 978-1-78139-289-8
Cover Photo: José Manuel Suárez
© 2012 Benediction Classics, Oxford.

Transcriber's Notes:

All apparent printer's errors retained. Variation in punctuation are as in the original, but missing full stops at end of paragraphs have been supplied. There are inconsistencies in the use of italics, spacing of words and use of full stop after 'AXIOME', abbreviations etc. All are retained to match text. There is a great variation in spelling including multiple spellings of the same word, all spelling has been retained to match text. There are several instances of obviously missing letters or inverted n & u. These have been changed or obvious letters replaced, with the changes surrounded by {}.

There are a number of instances in the original text where 'that' is immediately followed by a second 'that' in the sentence. These could be potential printer's errors or, since several of them make sense, part of the author's style. They have been left in the text as they appear in the original text.

The images have been retouched to clean up the diagrams and to improve readability of lettering where possible.

onsidering (Most Serene Prince) that the publishing this present Treatise, of so different an Argument from that which many expect, and which according to the intentions I proposed in my[1] Astronomicall *Adviso*, I should before this time have put forth, might peradventure make some thinke, either that I had wholly relinquished my farther imployment about the new Celestiall Observations, or that, at least, I handled them very remissely; I have judged fit to render an account, aswell of my deferring that, as of my writing, and publishing this treatise.

The Author's Observations of the Solar Spots

As to the first, the last discoveries of *Saturn* to be tricorporeall, and of the mutations of Figure in *Venus*, like to those that are seen in the Moon, together with the Consequents depending thereupon, have not so much occasioned the demur, as the investigation of the times of the Conversions of each of the Four Medicean Planets about *Jupiter*, which I lighted upon in *April* the year past, 1611, at my being in *Rome*; where, in the end, I assertained my selfe, that the first and neerest to *Jupiter*, moved about 8 *gr.* & 29 *m.* of its Sphere in an houre, makeing its whole revolution in one naturall day, and 18 hours, and almost an halfe. The second moves in its Orbe 14 *gr.* 13 *min.* or very neer, in an hour, and its compleat conversion is consummate in 3 dayes, 13 hours, and one third, or thereabouts. The third passeth in an hour, 2 *gr.* 6 *min.* little more or less of its Circle, and measures it all in 7 dayes, 4 hours, or very neer. The fourth, and more remote than the

---

[1] His Nuncio Siderio.

rest, goes in one houre, 0 *gr* 54 *min.* and almost an halfe of its Sphere, and finisheth it all in 16 dayes, and very neer 18 hours. But because the excessive velocity of their returns or restitutions, requires a most scrupulous precisenesse to calculate their places, in times past and future, especially if the time be for many Moneths or Years; I am therefore forced, with other Observations, and more exact than the former, and in times more remote from one another, to correct the Tables of such Motions, and limit them even to the shortest moment: for such exactnesse my first Observations suffice not; not only in regard of the short intervals of Time, but because I had not as then found out a way to measure the distances between the said Planets by any Instrument: I Observed such Intervals with simple relation to the Diameter of the Body of *Jupiter*; taken, as we have said, by the eye, the which, though they admit not errors of above a Minute, yet they suffice not for the determination of the exact greatness of the Spheres of those Stars. But now that I have hit upon a way of taking such measures without failing, scarce in a very few Seconds, I will continue the observation to the very occultation of *JUPITER*, which shall serve to bring us to the perfect knowledge of the Motions, and Magnitudes of the Orbes of the said Planets, together also with some other consequences thence arising. I adde to these things the observation of some obscure Spots, which are discovered in the Solar Body, which changing, position in that, propounds to our consideration a great argument either that the Sun revolves in it selfe, or that perhaps other Starrs, in like manner as *Venus* and *Mercury*, revolve about it, invisible in other times, by reason of their small digressions, lesse than that of *Mercury*, and only visible when they interpose between the Sun and our eye, or else hint the truth of both this and that; the certainty of which things ought not to be contemned, nor omitted.

Continuall observation hath at last assured me that these Spots are matters contiguous to the Body of the Sun, there continually produced in great number, and afterwards dissolved, some in a shorter, some in a longer time, and to be by the Conversion or Revolution of the Sun in it selfe, which in a Lunar Moneth, or thereabouts, finisheth its Period, caried about in a Circle, an accident great of it selfe, and greater for its Consequences.

GALILEI

The occasion inducing the Author to write this Treatise.

As to the other particular in the next place Many causes have moved me to write the present Tract, the subject whereof, is the Dispute which I held some dayes since, with some learned men of this City, about which, as your Highnesse knows, have followed many Discourses: The principall of which Causes hath been the Intimation of your Highnesse, having commended to me Writing, as a singular means to make true known from false, reall from apparent Reasons, farr better than by Disputing vocally, where the one or the other, or very often both the Disputants, through too greate heate, or exalting of the voyce, either are not understood, or else being transported by ostentation of not yeilding to one another, farr from the first Proposition, with the novelty, of the various Proposals, confound both themselves and their Auditors.

Moreover, it seemed to me convenient to informe your Highnesse of all the sequell, concerning the Controversie of which I treat, as it hath been advertised often already by others: and because the Doctrine which I follow, in the discussion of the point in hand, is different from that of *Aristotle*; and interferes with his Principles, I have considered that against the Authority of that most famous Man, which amongst many makes all suspected that comes not from the Schooles of the Peripateticks, its farr better to give ones Reasons by the Pen than by word of mouth, and therfore I resolved to write the present discourse: in which yet I hope to demonstrate that it was not out of capritiousnesse, or for that I had not read or understood *Aristotle*, that I sometimes swerve from his opinion, but because severall Reasons perswade me to it, and the same *Aristotle* hath tought me to fix my judgment on that which is grounded upon Reason, and not on the bare Authority of the Master; and it is most certaine according to the sentence of *Alcinoos*, that philosophating should be free. Nor is the resolution of our Question in my judgment without some benefit to the The benefit of this Argument. Universall, forasmuch as treating whether the figure of Solids operates, or not, in their going, or not going to the bottome in Water, in occurrences of building Bridges or other Fabricks on the Water, which happen commonly in affairs of grand import, it may be of great availe to know the truth.

I say therfore, that being the last Summer in company with certain Learned men, it was said in the argumentation; That Condensation was the propriety of Cold[2], and there was alledged for instance, the example of Ice: now I at that time said, that, in my judgment, the Ice should be rather Water rarified than condensed, and my reason was, because Condensation begets diminution of Mass, and augmentation of gravity, and Rarifaction causeth greater Lightness, and augmentarion of Masse: and Water in freezing, encreaseth in Masse, and the Ice made thereby is lighter than the Water on which it swimmeth.

What I say, is manifest, because, the medium subtracting from the whole Gravity of Sollids the weight of such another Masse of the said Medium; as Archimedes *proves in his[3]* First Booke De Insidentibus Humido; *when ever the Masse of the said Solid encreaseth by Distraction, the more shall the* Medium *detract from its entire Gravity; and lesse, when by Compression it shall be condensed and reduced to a lesse Masse.*

Figure operates not in the Natation of Sollids.

It was answered me, tha{t} that proceeded not from the greater Levity, but from the Figure, large and flat, which not being able to penetrate the Resistance of the Water, is the cause that it submergeth not. I replied, that any piece of Ice, of whatsoever Figure, swims upon the Water, a manifest signe, that its being never so flat and broad, hath not any part in its floating: and added, that it was a manifest proofe hereof to see a piece of Ice of very broad Figure being thrust to the botome of the Water, suddenly return to flote atoppe, which had it been more grave, and had its swimming proceeded from its Forme, unable to penetrate the Resistance of the *Medium*, that would be altogether impossible; I concluded therefore, that the Figure was in sort a Cause of the Natation or Submersion of Bodies, but the greater or lesse Gravity in respect of the Water: and therefore all Bodyes heavier than it of what Figure soever they be, indifferently go to the bottome, and the lighter, though of any figure, float indifferently on the top: and I suppose that those which hold otherwise, were induced to that beliefe, by seeing how that diversity of Formes or Figures, greatly altereth the Velosity, and Tardity of Motion; so that Bodies of Figure broad and thin, descend far more leasurely into the Water, than those of a more compacted Figure, though both made of the same Matter: by which

---

[2] According to the Peripateticks.
[3] *In lib: 1. of Natation of Bodies Prop. 7.*

some might be induced to believe that the Dilatation of the Figure might reduce it to such amplenesse that it should not only retard but wholly impede and take away the Motion, which I hold to be false. Upon this Conclusion, in many dayes discourse, was spoken much, and many things, and divers Experiments produced, of which your Highnesse heard, and saw some, and in this discourse shall have all that which hath been produced against my Assertion, and what hath been suggested to my thoughts on this matter, and for confirmation of my Conclusion: which if it shall suffice to remove that (as I esteem hitherto false) Opinion, I shall thinke I have not unprofitably spent my paynes and time. and although that come not to passe, yet ought I to promise another benefit to my selfe, namely, of attaining the knowledge of the truth, by hearing my Fallacyes confuted, and true demonstrations produced by those of the contrary opinion.

And to proceed with the greatest plainnesse and perspicuity that I can possible, it is, I conceive, necessary, first of all to declare what is the true, intrinsecall, and totall Cause, of the ascending of some Sollid Bodyes in the Water, and therein floating; or on the contrary, of their sinking and so much the rather in asmuch as I cannot satisfie myselfe in that which *Aristotle* hath left written on this Subject.

The cause of the Natation & submersion of Solids in the Water.

I say then the Cause why some Sollid Bodyes descend to the Bottom of Water, is the excesse of their Gravity, above the Gravity of the Water; and on the contrary, the excess of the Waters Gravity above the Gravity of those, is the Cause that others do not descend, rather that they rise from the Bottom, and ascend to the Surface. This was subtilly demonstrated by *Archimedes* in his Book Of the Natation of Bodies: Conferred afterwards by a very grave Author, but, if I erre not invisibly, as below for defence of him, I shall endeavour to prove.

I, with a different Method, and by other meanes, will endeavour to demonstrate the same, reducing the Causes of such Effects to more intrinsecall and immediate Principles, in which also are discovered the Causes of some admirable and almost incredible Accidents, as that would be, that a very little quantity of Water, should be able, with its small weight, to raise and sustain a Solid Body, an hundred or a thousand times heavier than it.

6

And because demonstrative Order so requires, I shall define certain Termes, and afterwards explain some Propositions, of which, as of things true and obvious, I may make use of to my present purpose.

DEFINITION I.

I then call equally Grave in specie, *those Matters of which equall Masses weigh equally.*

As if for example, two Balls, one of Wax, and the other of some Wood of equall Masse, were also equall in Weight, we say, that such Wood, and the Wax are *in specie* equally grave.

DEFINITION II.

But equally grave in Absolute Gravity, we call two Sollids, weighing equally, though of Mass they be unequall.

As for example, a Mass of Lead, and another of Wood, that weigh each ten pounds, I call equall in Absolute Gravity, though the Mass of the Wood be much greater then that of the Lead.

And, consequently, less Grave in specie.

DEFINITION III.

I call a Matter more Grave in specie *than another, of which a Mass, equall to a Mass of the other, shall weigh more.*

And so I say, that Lead is more grave *in specie* than Tinn, because if you take of them two equall Masses, that of the Lead weigheth more.

DEFINITION IV.

But I call that Body more grave absolutely than this, if that weigh more than this, without any respect had to the Masses.

And thus a great piece of Wood is said to weigh more than a little lump of Lead, though the Lead be *in specie* more heavy than the

Wood. And the same is to be understood of the less grave *in specie*, and the less grave absolutely.

These Termes defined, I take from the Mechanicks two Principles: the first is, that

AXIOME. I.

Weights absolutely equall, moved with equall Velocity, are of equall Force and Moment in their operations.

DEFINITION V.

Moment, amongst Mechanicians, signifieth that Vertue, that Force, or that Efficacy, with which the Mover moves, and the Moveable resists.

Which Vertue dependes not only on the simple Gravity, but on the Velocity of the Motion, and on the diverse Inclinations of the Spaces along which the Motion is made: For a descending Weight makes a greater Impetus in a Space much declining, than in one less declining; and in summe, what ever is the occasion of such Vertue, it ever retaines the name of Moment; nor in my Judgement, is this sence new in our Idiome, for, if I mistake not, I think we often say; This is a weighty businesse, but the other is of small moment: and we consider lighter matters and let pass those of Moment; a Metaphor, I suppose, taken from the Mechanicks.

As for example, two weights equall in absolute Gravity, being put into a Ballance of equall Arms, they stand in *Equilibrium*, neither one going down, nor the other up: because the equality of the Distances of both, from the Centre on which the Ballance is supported, and about which it moves, causeth that those weights, the said Ballance moving, shall in the same Time move equall Spaces, that is, shall move with equall Velocity, so that there is no reason for which this Weight should descend more than that, or that more than this; and therefore they make an *Equilibrium*, and their Moments continue of semblable and equall Vertue.

The second Principle is; That

AXIOME II.

The Moment and Force of the Gravity, is encreased by the Velocity of the Motion.

So that Weights absolutely equall, but conjoyned with Velocity unequall, are of Force, Moment and Vertue unequall: and the more potent, the more swift, according to the proportion of the Velocity of the one, to the Velocity of the other. Of this we have a very pertinent example in the Balance or Stiliard of unequall Arms, at which Weights absolutely equall being suspended, they do not weigh down, and gravitate equally, but that which is at a greater distance from the Centre, about which the Beam moves, descends, raising the other, and the Motion of this which ascends is slow, and the other swift: and such is the Force and Vertue, which from the Velocity of the Mover, is conferred on the Moveable, which receives it, that it can exquisitely compensate, as much more Weight added to the other slower Moveable: so that if of the Arms of the Balance, one were ten times as long as the other, whereupon in the Beames moving about the Centre, the end of that would go ten times as far as the end of this, a Weight suspended at the greater distance, may sustain and poyse another ten times more grave absolutely than it: and that because the Stiliard moving, the lesser Weight shall move ten times faster than the bigger. It ought always therefore to be understood, that Motions are according to the same Inclinations, namely, that if one of the Moveables move perpendicularly to the Horizon, then the other makes its Motion by the like Perpendicular; and if the Motion of one were to be made Horizontally; that then the other is made along the same Horizontall plain: and in summe, always both in like Inclinations. This proportion between the Gravity and Velocity is found in all Mechanicall Instruments: and is considered by *Aristotle*, as a Principle in his *Mechanicall Questions*; whereupon we also may take it for a true Assumption, That

AXIOME III.

Weights absolutely unequall, do alternately counterpoyse and become of equall Moments, as oft as their Gravities, with contrary proportion, answer to the Velocity of their Motions.

9

That is to say, that by how much the one is less grave than the other, by so much is it in a constitution of moving more swiftly than that.

Having prefatically explicated these things, we may begin to enquire, what Bodyes those are which totally submerge in Water, and go to the Bottom, and which those that by constraint float on the top, so that being thrust by violence under Water, they return to swim, with one part of their Mass visible above the Surface of the Water: and this we will do by considering the respective operation of the said Solids, and of Water: Which operation followes the Submersion and sinking; and this it is, That in the Submersion that the Solid maketh, being depressed downwards by its proper Gravity, it comes to drive away the water from the place where it successively subenters, and the water repulsed riseth and ascends above its first levell, to which Ascent on the other side it, as being a grave Body of its own nature, resists: And because the descending Solid more and more immerging, greater and greater quantity of Water ascends, till the whole Sollid be submerged; its necessary to compare the Moments of the Resistance of the water to Ascension, with the Moments of the pressive Gravity of the Solid: And if the Moments of the Resistance of the water, shall equalize the Moments of the Solid, before its totall Immersion; in this case doubtless there shall be made an *Equilibrium*, nor shall the Body sink any farther. But if the Moment of the Solid, shall alwayes exceed the Moments wherewith the repulsed water successively makes Resistance, that Solid shall not only wholly submerge under water, but shall descend to the Bottom. But if, lastly, in the instant of totall Submersion, the equality shall be made between the Moments of the prement Solid, and the resisting Water; then shall rest, ensue, and the said Solid shall be able to rest indifferently, in whatsoever part of the water. By this time is manifest the necessity of comparing the Gravity of the water, and of the Solid; and this comparison might at first sight seem sufficient to conclude and determine which are the Solids that float a-top, and which those that sink to the Bottom in the water, asserting that those shall float which are lesse grave *in specie* than the water, and those submerge, which are *in specie* more grave. For it seems in appearance, that the Sollid in sinking continually, raiseth so much Water in Mass, as answers to the parts of its own Bulk submerged: whereupon it is impossible, that a Solid less grave *in specie*, than water, should wholly sink, as being unable to raise a weight greater than its own, and such would a Mass of water equall to its own Mass be. And likewise it seems necessary, that the graver Solids do go to the Bottom, as being of a Force more than sufficient for the raising a Masse of water, equall to its own, though inferiour in weight. Nevertheless the business succeeds otherwise: and though the Conclusions are true, yet are the Causes thus assigned deficient, nor is it true, that the Solid in submerging, raiseth and repulseth Masses of Water, equall to the parts of it self submerged; but the Water repulsed, is alwayes less than the parts of the Solid submerged: and so much the more by how much the Vessell in which the Water is contained is narrower: in such manner that it hinders not, but that a Solid may submerge all under Water, without raising so much Water in Mass, as would equall the tenth or twentieth part of its own Bulk: like as on the contrary, a very small quantity of Water, may raise a very great Solid Mass, though such Solid should weigh absolutely a hundred times as much, or more, than the said Water, if so be that the Matter of that same Solid be *in specie* less grave than the Water. And thus a great Beam, as suppose of a 1000 weight, may be raised and born afloat by Water, which weighs not 50: and this happens when the Moment of the Water is compensated by the Velocity of its Motion.

But because such things, propounded thus in abstract, are somewhat difficult to be comprehended, it would be good to demonstrate them by particular examples; and for facility of demonstration, we will suppose the Vessels in which we are to put the Water, and place the Solids, to be inviron'd and included with sides erected perpendicular to the Plane of the Horizon, and the Solid that is to be put into such vessell to be either a streight Cylinder, or else an upright Prisme.

The which proposed and declared, I proceed to demonstrate the truth of what hath been hinted, forming the ensuing Theoreme.

THEOREME I.

The Mass of the Water which ascends in the submerging of a Solid, Prisme or Cylinder, or that abaseth in taking it out, is less than the Mass of the said Solid, so depressed or advanced: and hath to it the same proportion, that the Surface of the Water circumfusing the Solid, hath to the same circumfused Surface, together with the Base of the Solid.

The Proportion of the water raised to the Solid submerged.

Let *the Vessell be A B C D, and in it the Water raised up to the Levell E F G, before the Solid Prisme H I K be therein immerged; but after that it is depressed under Water, let the Water be raised as high as the Levell L M, the Solid H I K shall then be all under Water, and the Mass of the elevated Water shall be L G, which is less than*

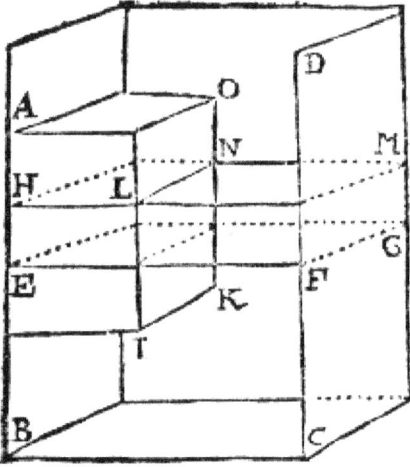

the Masse of the Solid depressed, namely of H I K, being equall to the only part E I K, which is con- tained under the first Levell E F G. Which is manifest, because if the Solid H I K be taken out, the Water I G shall return into the place oc- cupied by the Mass E I K, where it was continuate before the submersion of the Prisme. And the Mass L G being equall to the Mass E K: adde thereto the Mass E N, and it shall be the whole Mass E M, composed of the parts of the Prisme E N, and of the Water N F, equall to the whole Solid H I K: And, therefore, the Mass L G shall have the same proportion to E M, as to the Mass H I K: But the Mass L G hath the same proportion to the Mass E M, as the Surface L M hath to the Surface M H: Therefore it is manifest, that the Mass of Water repulsed L G, is in proportion to the Mass of the Solid submerged H I K; as the Surface L M, namely, that of the Water ambient about the Sollid, to the whole Surface H M, compounded of the said ambient water, and the Base of the Prisme H N. But if we suppose the first Levell of the Water the according to the Surface H M, and the Prisme allready submerged H I K; and after to be taken out and raised to E A O, and the Water to be faln from the first Levell H L M as low as E F G; It is manifest, that the Prisme E A O being the same with H I K, its superiour part H O, shall be equall to the inferiour E I K: and remove the common part E N, and, consequently, the Mass of the Water L G is equall to the Mass H O; and, therefore, less than the Solid, which is without the Water, namely, the whole Prisme E A O, to which likewise, the said Mass of Water abated L G, hath the same proportion, that the Surface of the Waters circumfused L M hath to the same circumfused Surface, together with the Base of the Prisme A O: which hath the same demonstration with the former case above.

And from hence is inferred, that the Mass of the Water, that riseth in the immersion of the Solid, or that ebbeth in elevating it, is not equall to all the Mass of the Solid, which is submerged or elevated, but to that part only, which in the immersion is under the first Levell of the Water, and in the elevation remaines above the first Levell: Which is that which was to be demonstrated. We will now pursue the things that remain.

And first we will demonstrate that,

THEOREME II.

When in one of the above said Vessels, of what ever breadth, whether wide or narrow, there is placed such a Prisme or Cylinder, inviron'd with Water, if we elevate that Solid perpendicularly, the Water circumfused shall abate, and the Abatement of the Water, shall have the same proportion to the Elevation of the Prisme, as one of the Bases of the Prisme, hath to the Surface of the Water Circumfused.

The proportion of the water abated, to the Solid raised.

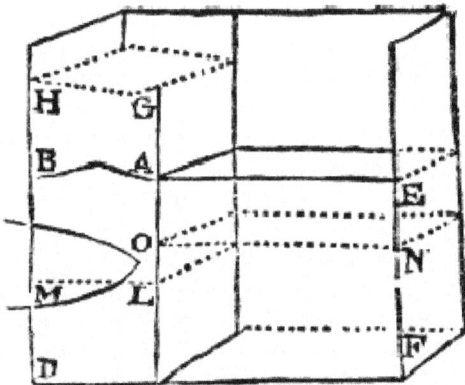

Imagine in the Vessell, as is aforesaid, the Prisme A C D B to be placed, and in the rest of the Space the Water to be diffused as far as the Levell E A: and raising the Solid, let it be transferred to G M, and let the Water be abased from E A to N O: I say, that the descent of the Water, measured by the Line A O, hath the same proportion to the rise of the Prisme, measured by the Line G A, as the Base of the Solid G H hath to the Surface of the Water N O. The which is manifest: because the Mass of the Solid G A B H, raised above the first Levell E A B, is equall to the Mass of Water that is abased E N O A. Therefore, E N O A and G A B H are two equall Prismes; for of equall Prismes, the Bases answer contrarily to their heights: Therefore, as the Altitude A O is to the Altitude A G, so is the Superficies or Base G H to the Surface of the Water N O. If therefore, for example, a Pillar were erected in a waste Pond full of Water, or else in a Well, capable of little more then the Mass of the said Pillar, in elevating the said Pillar, and taking it out of the Water, according as it riseth, the Water that invirons it will gradually abate, and the abasement of the Water at the instant of lifting out the Pillar, shall have the same proportion, that the thickness of the Pillar hath to the excess of the breadth of the said Pond or Well, above the thickness of the said Pillar: so that if the breadth of the Well were an eighth part larger than the thickness of the Pillar, and the breadth of the Pond twenty five times as great as the said thickness, in the Pillars ascending one foot, the water in the Well shall descend seven foot, and that in the Pond only 1/25 of a foot.

Why a Solid less grave *in specie* than water, stayeth not under water, in very small depths:

This Demonstrated, it will not be difficult to show the true cause, how it comes to pass, that,

THEOREME III.

A Prisme or regular Cylinder, of a substance specifically less grave than Water, if it should be totally submerged in Water, stayes not underneath, but riseth, though the Water circumfused be very little, and in absolute Gravity, never so much inferiour to the Gravity of the said Prisme.

Let then the Prisme A E F B, be put into the Vessell C D F B, the same being less grave *in specie* than the Water: and let the Water infused rise to the height of the Prisme: I say, that the Prisme left at liberty, it shall rise, being born up by the Water circumfused C D E

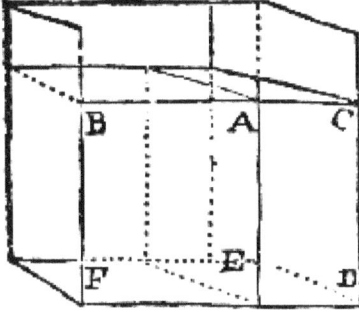

A. For the Water C E being specifically more grave than the Solid A F, the absolute weight of the water C E, shall have greater proportion to the absolute weight of the Prisme A F, than the Mass C E hath to the Mass A F (in regard the Mass hath the same proportion to the Mass, that the weight absolute hath to the weight absolute, in case the Masses are of the same Gravity *in specie*.) But the Mass C E is to the Mass A F, as the Surface of the water A C, is to the Superficies, or Base of the Prisme A B; which is the same proportion as the ascent of the Prisme when it riseth, hath to the descent of the Water circumfused C E.

Therefore, the absolute Gravity of the water C E, hath greater proportion to the absolute Gravity of the Prisme A F; than the Ascent of the Prisme A F, hath to the descent of the said water C E. The Moment, therefore, compounded of the absolute Gravity of the water C E, and of the Velocity of its descent, whilst it forceably repulseth and raiseth the Solid A F, is greater than the Moment compounded of the absolute Gravity of the Prisme A F, and of the Tardity of its ascent, with which Moment it contrasts and resists the repulse and violence done it by the Moment of the water: Therefore, the Prisme shall be raised.

It followes, now, that we proceed forward to demonstrate more particularly, how much such Solids shall be inferiour in Gravity to the water elevated; namely, what part of them shall rest submerged, and what shall be visible above the Surface of the water: but first it is necessary to demonstrate the subsequent Lemma.

The Proportion according to which the Submersion & Natation of Solids is made.

LEMMA I.

The absolute Gravities of Solids, have a proportion compounded of the proportions of their specificall Gravities, and of their Masses.

Let A and B be two Solids. I say, that the Absolute Gravity of A, hath to the Absolute Gravity of B, a proportion compounded of the proportions of the specificall Gravity of A, to the Specificall

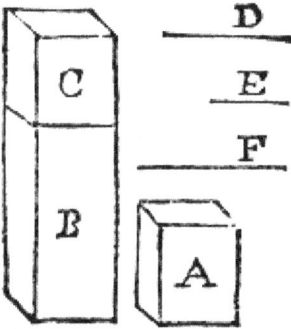

Gravity of B, and of the Mass A to the Mass B. Let the Line D have the same proportion to E, that the specifick Gravity of A, hath to the specifick Gravity of B; and let E be to F, as the Mass A to the Mass B: It is manifest, that the proportion of D to F, is compounded of the proportions D and E; and E and F. It is requisite, therefore, to demonstrate, that as D is to F, so the absolute Gravity of A, is to the absolute Gravity of B. Take the Solid C, equall in Mass to the Solid A, and of the same Gravity *in specie* with the Solid B. Because, therefore, A and C are equall in Mass, the absolute Gravity of A, shall have to the absolute Gravity of C, the same proportion, as the specificall Gravity of A, hath to the specificall Gravity of C, or of B, which is the same *in specie*; that is, as D is to E. And, because, C and B are of the same Gravity *in specie*, it shall be, that as the absolute weight of C, is to the absolute weight of B, so the Mass C, or the Mass A, is to the Mass B; that is, as the Line E to the Line F. As therefore, the absolute Gravity of A, is to the absolute Gravity of C, so is the Line D to the Line E: and, as the absolute Gravity of C, is to the absolute Gravity of B, so is the Line E to the Line F: Therefore, by Equality of proportion, the absolute Gravity of A, is to the absolute

Gravity of B, as the Line D to the Line F: which was to be demonstrated. I proceed now to demonstrate, how that,

THEOREME IV.

If a Solid, Cylinder, or Prisme, lesse grave specifically than the Water, being put into a Vessel, as above, of whatsoever greatnesse, and the Water, be afterwards infused, the Solid shall rest in the bottom, unraised, till the Water arrive to that part of the Altitude, of the said Prisme, to which its whole Altitude hath the same proportion, that the Specificall Gravity of the Water, hath to the Specificall Gravity of the said Solid: but infusing more Water, the Solid shall ascend.

Let the Vessell be M L G N of any bigness, and let there be placed in it the Solid Prisme D F G E, less grave *in specie* than the water; and look what proportion the Specificall Gravity of the water, hath to that of the Prisme, such let the Altitude D F, have to the Altitude F B. I say, that infusing water to the Altitude F B, the Solid D G shall not float, but shall stand in *Equilibrium*, so, that that every little quantity of water, that is infused, shall raise it. Let the water, therefore, be infused to the Levell A B C; and; because the Specifick Gravity of the Solid D G, is to the Specifick Gravity of the water, as the altitude B F is to the altitude F D; that is, as the Mass B G to the Mass G D; as the proportion of the Mass B G is to the Mass G D, as the proportion of the Mass G D is to the Mass A F, they compose the Proportion of the Mass B G to the Mass A F. Therefore, the Mass B G is to the Mass A F, in a proportion compounded of the proportions of the Specifick

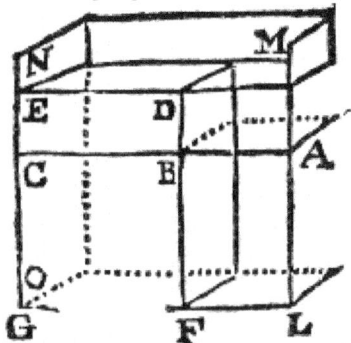

Gravity of the Solid G D, to the Specifick Gravity of the water, and of the Mass G D to the Mass A F: But the same proportions of the Specifick Gravity of G D, to the Specifick Gravity of the water, and of the Mass G D to the Mass A F, do also by

the precedent *Lemma*, compound the proportion of the absolute Gravity of the Solid D G, to the absolute Gravity of the Mass of the water A F: Therefore, as the Mass B G is to the Mass A F, so is the Absolute Gravity of the Solid D G, to the Absolute Gravity of the Mass of the water A F. But as the Mass B G is to the Mass A F; so is the Base of the Prisme D E, to the Surface of the water A B; and so is the descent of the water A B, to the Elevation of the Prisme D G; Therefore, the descent of the water is to the elevation of the Prisme, as the absolute Gravity of the Prisme, is to the absolute Gravity of the water: Therefore, the Moment resulting from the absolute Gravity of the water A F, and the Velocity of the Motion of declination, with which Moment it forceth the Prisme D G, to rise and ascend, is equall to the Moment that results from the absolute Gravity of the Prisme D G, and from the Velocity of the Motion, wherewith being raised, it would ascend: with which Moment it resists its being raised: because, therefore, such Moments are equall, there shall be an *Equilibrium* between the water and the Solid. And, it is manifest, that putting a little more water unto the other A F, it will increase the Gravity and Moment, whereupon the Prisme D G, shall be overcome, and elevated till that the only part B F remaines submerged. Which is that that was to be demonstrated.

COROLLARY I.

By what hath been demonstrated, it is manifest, that Solids less grave in specie *than the water, submerge only so far, that as much water in Mass, as is the part of the Solid submerged, doth weigh absolutely as much as the whole Solid.*

For, it being supposed, that the Specificall Gravity of the water, is to the Specificall Gravity of the Prisme D G, as the Altitude D F, is to the Altitude F B; that is, as the Solid D G is to the Solid B G; we might easily demonstrate, that as much water in Mass as is equall to the Solid B G, doth weigh absolutely as much as the whole Solid D G; For, by the *Lemma* foregoing, the Absolute Gravity of a Mass of water, equall to the Mass B G, hath to the Absolute Gravity of the Prisme D G, a proportion compounded of the proportions, of the Mass B G to the Mass G D, and of the Specifick Gravit{y} of the water, to the Specifick Gravity of the Prisme: But the Gravity *in specie* of the water, to the Gravity *in specie* of the Prisme, is supposed to be as the Mass G D to the Mass G B. Therefore, the Absolute Gravity of a Mass of water, equall to the Mass B G, is to the Absolute Gravity of the Solid D G, in a proportion compounded of the proportions, of the Mass B G to the Mass G D, and of the Mass D G to the Mass G B; which is a proportion of equalitie. The Absolute Gravity, therefore, of a Mass of Water equall to the part of the Mass of the Prisme B G, is equall to the Absolute Gravity of the whole Solid D G.

COROLLARY II.

It followes, moreover, that a Solid less grave than the water, being put into a Vessell of any imaginable greatness, and water being circumfused about it to such a height, that as much water in Mass, as is the part of the Solid submerged, do weigh absolutely as much as the whole Solid; it shall by that water be justly sustained, be the circumfused Water in quantity greater or lesser.

A Rule to equilibrate Solids in the water.

For, if the Cylinder or Prisme M, less grave than the water, *v. gra.* in Subsequiteriall proportion, shall be put into the capacious Vessell A B C D, and the water raised about it, to three quarters of its

height, namely, to its Levell A D: it shall be sustained and exactly poysed in *Equilibrium*. The same will happen; if the Vessell E N S F were very small, so, that between the Vessell and the Solid M, there were but a very narrow space, and only capable of so much water, as the hundredth part of the Mass M, by which it should be likewise raised and erected, as before it had been elevated to three fourths of the height of the Solid: which to many at the first sight, may seem a notable Paradox, and beget a conceit, that the Demonstration of these effects, were sophisticall and fallacious: but, for those who so repute it, the Experiment is a means that may fully satisfie them. But he that shall but comprehend of what Importance Velocity of Motion is, and how it exactly compensates the defect and want of Gravity, will cease to wonder, in considering that at the elevation of the Solid M, the great Mass of water A B C D abateth very little, but the little Mass of water E N S F decreaseth very much, and in an instant, as the Solid M before

19

did rise, howbeit for a very short space: Whereupon the Moment, compounded of the small Absolute Gravity of the water E N S F, and of its great Velocity in ebbing, equalizeth the Force and and Moment, that results from the composition of the immense Gravity of the water A B C D, with its great slownesse of ebbing; since that in the Elevation of the Sollid M, the abasement of the lesser water E S, is performed just so much more swiftly than the great Mass of water A C, as this is more in Mass than that which we thus demonstrate.

In the rising of the Solid M, its elevation hath the same proportion to the circumfused water E N S F, that the Surface of the said water, hath to the Superficies or Base of the said Solid M; which Base hath the same proportion to the Surface of the water A D, that the abasement or ebbing of the water A C, hath to the rise or elevation of the said Solid M. Therefore, by Perturbation of proportion, in the ascent of the said Solid M, the abasement of the water A B C D, to the abasement of the water E N S F, hath the same proportion, that the Surface of the water E F, hath to the Surface of the water A D; that is, that the whole Mass of the water E N S F, hath to the whole Mass A B C D, being equally high: It is manifest, therefore, that in the expulsion and elevation of the Solid M, the water E N S F shall exceed in Velocity of *M*otion the water A B C D, asmuch as it on the other side is exceeded by that in quantity: whereupon their Moments in such operations, are mutually equall.

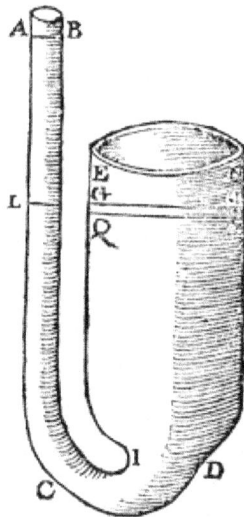

And, for ampler confirmation, and clearer explication of this, let us consider the present Figure, (which if I be not deceived, may serve to detect the errors of some Practick Mechanitians who upon a false foundation some times attempt impossible enterprizes,) in which, unto the large Vessell E I D F, the narrow Funnell or Pipe I C A B is continued, and suppose water infused into them, unto the Levell L G H, which water shall rest in this position, not without admiration in some, who cannot conceive how it can be, that the heavie charge of the great Mass of water G D, pressing downwards, should not elevate and repulse the little quantity of the other, contained in the Funnell or Pipe C L, by which the descent of it is resisted and hindered: But such wonder shall cease, if we begin to suppose the water G D to be abased only to Q D, and shall afterwards consider, what the water C L hath done, which to give place to the other, which is descended from the Levell G H, to the Levell Q O, shall of necessity have ascended in the same time, from the Levell L unto A B. And the ascent L B, shall be so much greater than the descent G Q, by how much the breadth of the Vessell G D, is greater than that of the Funnell I C; which, in summe, is as much as the water G D, is more than the water L C: but in regard that the Moment of the Velocity of the Motion, in one Moveable, compensates that of the Gravity of another what wonder is it, if the swift ascent of the lesser Water C L, shall resist the slow descent of the greater G D?

The same, therefore, happens in this operation, as in rhe Stilliard, in which a weight of two pounds counterpoyseth an other of 200, asoften as that shall move in the same time, a space 100 times greater than this: which falleth out when one Arme of the Beam is an hundred times as long as the other. Let the erroneous opinion of those therefore cease, who hold that a Ship is better, and easier born up in A ship flotes as well in ten Tun of Water as in an Ocean. a great abundance of water, then in a lesser quantity, (*this was believed by* Aristotle *in his Problems, Sect. 23, Probl. 2.*) it being on the contrary true, that its possible, that a Ship may as well float in ten Tun of water, as in an Ocean.

A Solid specifiaclly graver than the water, cannot be born up by any quantity of it.

But following our matter, I say, that by what hath been hitherto demonstrated, we may understand how, that

21

COROLLARY III.

One of the above named Solids, when more grave in specie *than the water, can never be sustained, by any whatever quantity of it.*

For having seen how that the Moment wherewith such a Solid, as grave *in specie* as the water, contrasts with the Moment of any Mass of water whatsoever, is able to retain it, even to its totall Submersion, without its ever ascending; it remaineth, manifest, that the water is far less able to raise it up, when it exceeds the same *in specie*: so, that though you infuse water till its totall Submersion, it shall still stay at the Bottome, and with such Gravity, and Resistance to Elevation, as is the excess of its Absolute Gravity, above the Absolute Gravity of a Mass equall to it, made of water, or of a Matter *in specie* equally grave with the water: and, though you should moreover adde never so much water above the Levell of that which equalizeth the Altitude of the Solid, it shall not, for all that, encrease the Pression, or Gravitation, of the parts circumfused about the said Solid, by which greater pression, it might come to be repulsed; because, the Resistance is not made, but only by those parts of the water, which at the Motion of the said Solid do also move, and these are those only, which are comprehended by the two Superficies equidistant to the Horizon, and their parallels, that comprehend the Altitude of the Solid immerged in the water.

I conceive, I have by this time sufficiently declared and opened the way to the contemplation of the true, intrinsecall and proper Causes of diverse Motions, and of the Rest of many Solid Bodies in diverse *Mediums*, and particularly in the water, shewing how all in effect, depend on the mutuall excesses of the Gravity of the Moveables and of the *Mediums*: and, that which did highly import, removing the Objection, which peradventure would have begotten much doubting, and scruple in some, about the verity of my Conclusion, namely, how that notwithstanding, that the excess of the Gravity of the water, above the Gravity of the Solid, demitted into it, be the cause of its floating and rising from the Bottom to the Surface, yet a quantity of water, that weighs not ten pounds, can raise a Solid that weighs above 100 pounds: in that we have demonstrated, That it sufficeth, that such difference be found between the Specificall Gravities of the *Mediums* and Moveables, let the particular and absolute Gravities be what they will: insomuch, that a Solid, provided that it be Specifically less grave than the water, although its absolute weight were 1000 pounds, yet may it be born up and elevated by ten pounds of water, and less: and on the contrary, another Solid, so that it be Specifically more grave than the water, though in absolute Gravity it were not above a pound, yet all the water in the Sea, cannot raise it from the Bottom, or float it. This sufficeth me, for my present occasion, to have, by the above declared Examples, discovered and demonstrated, without extending such matters farther, and, as I might have done, into a long Treatise: yea, but that there was a necessity of resolving the above proposed doubt, I should have contented my self with that only, which is demonstrated by *Archimedes*, in his first *Book De Insidentibus humido*: where in generall termes he infers and confirms the same *Of Natation* (a) *Lib. 1, Prop. 4.* (b) *Id. Lib. 1. Prop. 3.* (c) *Id. Lib. 1. Prop. 3.* Conclusions, namely, that Solids (*a*) less grave than water, swim or float upon it, the (*b*) more grave go to the Bottom, and the (*c*) equally grave rest indifferently in all places, yea, though they should be wholly under water.

But, because that this Doctrine of Archimedes, perused, transcribed and examined by *Signor Francesco Buonamico*, in his *fifth Book of Motion, Chap. 29*, and afterwards by him confuted, might by the Authority of so renowned, and famous a Philosopher, be rendered dubious, and suspected of falsity; I have judged it necessary to defend it, if I am able so to do, and to clear *Archimedes*, from those censures, with which he appeareth to be charged. *Buonamico* rejecteth the Doctrine of *Archimedes*, first, as not consentaneous against the Doctrine of *Archimedes*. with the Opinion of *Aristotle*, adding, that it was a strange thing to him, that the Water should exceed the Earth in Gravity, seeing on the contrary, that the Gravity of water, increaseth, by means of the participation of Earth. And he subjoyns presently after, that he was not satisfied with the Reasons of *Archimedes*, as not being able with that Doctrine, to assign the cause whence it comes, that a Boat and a Vessell, which otherwise, floats above the water, doth sink to the Bottom, if once it be filled with water; that by reason of the equality of Gravity, between the water within it, and the other water without, it should stay a top; but yet, nevertheless, we see it to go to the Bottom.

He farther addes, that *Aristotle* had clearly confuted the Ancients, who said, that light Bodies moved upwards, driven by the impulse of the more grave Ambient: which if it were so, it should seem of necessity to follow, that all naturall Bodies are by nature heavy, and none light: For that the same would befall the Fire and Air, if put in the Bottom of the water. And, howbeit, *Aristotle* grants a Pulsion in the Elements, by which the Earth is reduced into a Sphericall Figure, yet nevertheless, in his judgement; it is not such that it can remove grave Bodies from their naturall places, but rather, that it send them toward the Centre, to which (as he somewhat obscurely continues to say,) the water principally moves, if it in the interim meet not with something that resists it, and, by its Gravity, thrusts it out of its place: in which case, if it cannot directly, yet at least as well as it can, it tends to the Centre: but it happens, that light Bodies by such Impulsion, do all ascend upward: but this properly they have by nature, as also, that other of swimming. He concludes, lastly, that he concurs with *Archimedes* in his Conclusions; but not in the Causes, which he would referre to the facile and difficult Separation of the *Medium*, and to the predominance of the Elements, so that when the Moveable superates the power of the *Medium*; as for example, Lead doth the Continuity of water, it shall move thorow it, else not.

This is all that I have been able to collect, as produced against *Archimedes* by *Signor Buonamico*: who hath not well observed the Principles and Suppositions of *Archimedes*; which yet must be false, if the Doctrine be false, which depends upon them; but is contented to alledge therein some Inconveniences, and some Repugnances to the Doctrine and Opinion of *Aristotle*. In answer to which Objections, I say, first, That the being of *Archimedes* Doctrine, simply different from the Doctrine of *Aristotle*, ought not to move any to suspect it, there being no cause, why the Authority of this should be preferred to the Authority of the other: but, because, where the decrees of Nature are indifferently exposed to the intellectuall eyes of each, the Authority of the one and the other, loseth all a{u}thenticalness of Perswasion, the absolute power residing in Reason; therefore I pass to that which he alledgeth in the second place, as an absurd consequent . of the Doctrine of *Archimedes*, namely, That water should be more grave than Earth. But I really find not, that ever *Archimedes* said such a thing, or that it can be rationally deduced from his Conclusions: and if that were manifest unto me, I verily believe, I should renounce his Doctrine, as most erroneous. Perhaps this Deduction of *Buonamico*, is founded upon that which he citeth of the Vessel, which swims as long as its voyd of water, but once full it sinks to the Bottom, and understanding it of a Vessel of Earth, he infers against *Archimedes* thus: Thou sayst that the Solids which swim, are less grave than water: this Vessell swimmeth: therefore, this Vessell is lesse grave than water. If this be the Illation. I easily answer, granting that this Vessell is lesse grave than water, and denying the other consequence, namely, that Earth is less Grave than Water. The Vessel that swims occupieth in the water, not only a place equall to the Mass of the Earth, of which it is formed; but equall to the Earth and to the Air together, contained in its concavity. And, if such a Mass compounded of Earth and Air, shall be less grave than such another quantity of water, it shall swim, and shall accord with the Doctrine of *Archimedes*; but if, again, removing the Air, the Vessell shall be filled with water, so that the Solid put in the water, be nothing but Earth, nor occupieth other place, than that which is only possest by Earth, it shall then go to the Bottom, by reason that the Earth is heavier than the water: and this corresponds well with the meaning of *Archimedes*. See the same effect illustrated, with such another Experiment, In pressing a Viall Glass to the Bottom of the water, when it is full of Air, it will meet with great resistance, because it is not the Glass alone, that is pressed under water, but together with the Glass a great Mass of Air, and such, that if you should take as much water, as the Mass of the Glass, and of the Air contained in it, you would have a weight much greater than that of the Viall, and of its Air: and, therefore, it will not submerge without great violence: but if we demit only the Glass into the water, which shall be when you shall fill the Glass with water, then shall the Glass descend to the Bottom; as superiour in Gravity to the water.

Returning, therefore, to our first purpose; I say, that Earth is more grave than water, and that therefore, a Solid of Earth goeth to the bottom of it; but one may possibly make a composition of Earth and Air, which shall be less grave than a like Mass of Water; and this shall swim: and yet both this and the other experiment shall very well accord with the Doctrine of *Archimedes*. But because that in my judgment it hath nothing of difficulty in it, I will not positively affirme that *Signor Buonamico*, would by such a discourse object unto *Archimedes* the absurdity of inferring by his doctrine, that Earth was less grave than Water, though I know not how to conceive what other accident he could have induced thence.

Perhaps such a Probleme (in my judgement false) was read by *Signor Buonamico* in some other Author, by whom peradventure it was attributed as a singular propertie, of some particular Water, and so comes now to be used with a double errour in confutation of *Ar-*

chimedes, since he saith no such thing, nor by him that did say it was it meant of the common Element of Water.

The third difficulty in the doctrine of *Archimedes* was, that he could not render a reason whence it arose, that a piece of Wood, and a Vessell of Wood, which otherwise floats, goeth to the bottom, if filled with Water. *Signor Buonamico* hath supposed that a Vessell of Wood, and of Wood that by nature swims, as before is said, goes to the bottom, if it be filled with water; of which he in the following Chapter, which is the 30 of the fifth Book copiously discourseth: but I (speaking alwayes without diminution of his singular Learning) dare in defence of *Archimedes* deny this experiment, being certain that a piece of Wood which by its nature sinks not in Water, shall not sinke though it be turned and converted into the forme of any Vessell whatsoever, and then filled with Water: and he that would readily see the Experiment in some other tractable Matter, and that is easily reduced into several Figures, may take pure Wax, and making it first into a Ball or other solid Figure, let him adde to it so much Lead as shall just carry it to the bottome, so that being a graine less it could not be able to sinke it, and making it afterwards into the forme of a Dish, and filling it with Water, he shall finde that without the said Lead it shall not sinke, and that with the Lead it shall descend with much slowness: & in short he shall satisfie himself, that the Water included makes no alteration. I say not all this while, but that its possible of Wood to make Barkes, which being filled with water, sinke; but that proceeds not through its Gravity, encreased by the Water, but rather from the Nailes and other Iron Workes, so that it no longer hath a Body less grave than Water, but one mixt of Iron and Wood, more grave than a like Masse of Water. Therefore let *Signor Buonamico* desist from desiring a reason of an effect, that is not in nature: yea if the sinking of the Woodden Vessell when its full of Water, may call in question the Doctrine of *Archimedes*, which he would not have you to follow, is on the contrary consonant and agreeable to the Doctrine of the Peripateticks, since it aptly assignes a reason why such a Vessell must, when its full of Water, descend to the bottom; converting the Argument the other way, we may with safety say that the Doctrine of *Archimedes* is true, since it aptly agreeth with true experiments, and question the other, whose Deductions are fastened upon erroneouss Conclusions. As for the other point hinted in this same Instance, where it seemes that *Benonamico* understands the same not only of a piece of wood, shaped in the forme of a Vessell, but also of massie Wood, which filled, *scilicet*, as I believe, he would say, soaked and steeped in Water, goes finally to the bottom that happens in some porose Woods, which, while their Porosity is replenished with Air, or other Matter less grave than Water, are Masses specificially less grave than the said Water, like as is that Viall of Glass whilest it is full of Air: but when, such light Matter departing, there succeedeth Water into the same Porosities and Cavities, there results a compound of Water and Glass more grave than a like Mass of Water: but the excess of its Gravity consists in the Matter of the Glass, and not in the Water, which cannot be graver than it self: so that which remaines of the Wood, the Air of its Cavities departing, if it shall be more grave *in specie* than Water, fil but its Porosities with Water, and you shall have a Compost of Water and of Wood more grave than Water, but not by vertue of the Water received into and imbibed by the Porosities, but of that Matter of the Wood which remains when the Air is departed: and being such it shall, according to the Doctrine of *Archimedes*, goe to the bottom, like as before, according to the same Doctrine it did swim.

As to that finally which presents itself in the fourth place, namely, that the *Ancients* have been heretofore confuted by *Aristotle*, who denying Positive and Absolute Levity, and truely esteeming all Bodies to be grave, said, that that which moved upward was driven by the circumambient Air, and therefore that also the Doctrine of *Archimedes*, as an adherent to such an Opinion was convicted and confuted: I answer first, that *Signor Buonamico* in my judgement hath imposed upon *Archimedes*, and deduced from his words more than ever he intended by them, or may from his Propositions be collected, in regard that *Archimedes* neither denies, nor admitteth Positive Levity, nor doth he so much as mention it: so that much less ought *Buonamico* to inferre, that he hath denied that it might be the Cause and Principle of the Ascension of Fire, and other Light Bodies: having but only demonstrated, that Solid Bodies more grave than Water descend in it, according to the excess of their Gravity above the Gravity of that, he demonstrates likewise, how the less grave ascend in the same Water, accordng to its excess of Gravity, above the Gravity of them. So that the most that can be gathered from the Demonstration of *Archimedes* is, that like as the excess of the Gravity of the Moveable above the Gravity of the Water, is the Cause that it descends therein, so the excess of the Gravity of the water above that of the Moveable, is a sufficient Cause why it descends not, but rather betakes it self to swim: not enquiring whether of moving upwards there is, or is not any other Cause contrary to Gravity: nor doth *Archimedes* discourse less properly than if one should say: If the South Winde shall assault the Barke with greater *Impetus* than is the violence with which the Streame of the River carries it towards the South, the motion of it shall be towards the North: but if the *Impetus* of the Water shall overcome that of the Winde, its motion shall be towards the South. The discourse is excellent and would be unworthily contradicted by such as should oppose it, saying: Thou mis-alledgest as Cause of the motion of the Bark towards the South, the *Impetus* of the Stream of the Water above that of the South Winde; mis-alledgest I say, for it is the Force of the North Winde opposite to the South, that is able to drive the Bark towards the South. Such an Objection would be superfluous, because he which alledgeth for Cause of the Motion the Stream of the Water, denies not but that the Winde opposite to the South may do the same, but only affirmeth that the force of the Water prevailing over the South Wind, the Bark shall move towards the South: and saith no more than is true. And just thus when *Archimedes* saith, that the Gravity of the Water prevailing over that by which the moveable descends to the Bottom, such moveable shall be raised from the Bottom to the Surface alledgeth a very true Cause of such an Accident, nor doth he affirm or deny that there is, or is not, a vertue contrary to Gravity, called by some Levity, that hath also a power of moving some Matters upwards. Let therefore the Weapons of *Signor Buonamico* be directed against *Plato*, and other *Ancients*, who totally denying *Levity*, and taking all Bodies to be grave, say that the Motion upwards is made, not from an intrinsecal Principle of the Moveable, but only by the Impulse of the *Medium*; and let *Archimedes* and his Doctrine escape him, since he hath given him no Cause of quarelling with him. But if this Apologie, produced in defence of *Archimedes*, should seem to some insufficient to free him from the Objections and Arguments, produced by *Aristotle* against *Plato*, and the other *Ancients*, as if they did also fight against *Archimedes*, alledging the Impulse of the Water as the Cause of the swimming of some Bodies less grave than it, I would not question, but that I should be able to maintaine the Doctrine of *Plato* and those others to be most true, who absolutely deny Levity, and affirm no other Intrinsecal Principle of Motion to be in Elementary Bodies save only that towards the Centre of the Earth, nor no other Cause of moving upwards, speaking of that which hath the resemblance of natural Motion, but only the repulse of the *Medium*, fluid, and exceeding the Gravity of the Moveable: and as to the Reasons of *Aristotle* on the contrary, I believe that I could be able fully to answer them, and I would assay to do it, if it were absolutely necessary to the present Matter, or were it not too long a Digression for this short Treatise. I will only say, that if there were in some of our Ellementary Bodies an Intrinsecall Principle and Naturall Inclination to shun the Centre of the Earth, and to move towards the Concave of the Moon, such Bodies, without doubt, would more swiftly ascend through those *Mediums* that least oppose the Velocity of the Moveable, and these are the more tenuous and subtle; as is, for example, the Air in comparison of the Water, we daily proving that we can with farre more expeditious Velocity move a Hand or a Board to and again in one than in the other: nevertheless, we never could finde any Body, that did not ascend much more swiftly in the water than in the Air. Yea of Bodies which we see continually to ascend in the Water, there is none that having arrived to the confines of the Air, do not wholly lose their Motion; even the Air it self, which rising with great Celerity through the Water, being once come to its Region it loseth all

And, howbeit, Experience shewes, that the Bodies, successively less grave, do most expeditiously ascend in water, it cannot be doubted, but that the Ignean Exhalations do ascend more swiftly through the water, than doth the Air: which Air is seen by Experience to ascend more swiftly through the Water, than the Fiery Exhalations through the Air: Therefore, we must of necessity conclude, that the said Exhalations do much more expeditiously ascend through the Water, than through the Air; and that, consequently, they are moved by the Impulse of the Ambient *Medium*, and not by an intrinsick Principle that is in them, of avoiding the Centre of the Earth; to which other grave Bodies tend.

The Authors confutation of the Peripateticks Causes of Natation & Submersion.

To that which for a finall conclusion, *Signor Buonamico* produceth of going about to reduce the descending or not descending, to the easie and uneasie Division of the *Medium*, and to the predominancy of the Elements: I answer, as to the first part, that that cannot in any manner be admitted as a Cause, being that in none of the Fluid *Mediums*, as the Air, the Water, and other Liquids, there is any Resistance against Division, but all by every the least Force, are divided and penetrated, as I will anon demonstrate: so, that of such Resistance of Division there can be no Act, since it self is not in being. As to the other part, I say, that the predominancy of the Elements in Moveables, is to be considered, as far as to the excesse or defect of Gravity, in relation to the *Medium*: for in that Action, the Elements operate not, but only, so far as they are grave or light: therefore, to say that the Wood of the Firre sinks not, because Air predominateth in it, is no more than to say, because it is less grave than the Water. Yea, even the immediate Cause, is its being less grave than the Water: and it being under the predominancy of the Air, is the Cause of its less Gravity: Therefore, he that alledgeth the predominancy of the Element for a Cause, brings the Cause of the Cause, and not the neerest and immediate Cause. Now, who knows not that the true Cause is the immediate, and not the mediate? Moreover, he that alledgeth Gravity, brings a Cause most perspicuous to Sence: The cause we may very easily assertain our selves; whether Ebony, for example, and Firre, be more or less grave than water: but whether Earth or Air predominates in them, who shall make that manifest? Certainly, no Experiment can better do it than to observe whether they swim or sink. So, that he who knows, not whether such a Solid swims, unless when he knows that Air predominates in it, knows not whether it swim, unless he sees it swim, for then he knows that it swims, when he knows that it is Air that predominates, but knows not that Air hath the predominance, unless he sees it swim: therefore, he knows not if it swims, till such time as he hath seen it swim.

Let us not then despise those Hints, though very dark, which Reason, after some contemplation, offereth to our Intelligence, and lets be content to be taught by *Archimedes*, that then any Body shall submerge in water, when it shall be specifically more grave than it, and that if it shall be less grave, it shall of necessity swim, and that it will rest indifferently in any place under water, if its Gravity be perfectly like to that of the water.

These things explained and proved, I come to consider that which offers it self, touching what the Diversity of figure given unto the said Moveable hath to do with these Motions and Rests; and proceed to affirme, that,

THEOREME V.

The diversity of Figures given to this or that Solid, cannot any way be a Cause of its absolute Sinking or Swimming.

So that if a Solid being formed, for example, into a Sphericall Figure, doth sink or swim in the water, I say, that being formed into any other Figure, the same figure in the same water, shall sink or swim: nor can such its Motion by the Expansion or by other mutation of Figure, be impeded or taken away.

The Expansion of Figure, retards the Velocity of the ascent or descent of the Moveable in the water; but doth not deprive it of all Motion.

The Expansion of the Figure may indeed retard its Velocity, aswell of ascent as descent, and more and more according as the said Figure is reduced to a greater breadth and thinness: but that it may be reduced to such a form as that that same matter be wholly hindred from moving in the same water, that I hold to be impossible. In this I have met with great contradictors, who producing some Experiments, and in perticular a thin Board of Ebony, and a Ball of the same Wood, and shewing how the Ball in Water descended to the bottom, and the Board being put lightly upon the Water submerged not, but rested; have held, and with the Authority of *Aristotle*, confirmed themselves in their Opinions, that the Cause of that Rest was the breadth of the Figure, u{n}able by its small weight to pierce and penetrate the Resistance of the Waters Crassitude, which Resistance is readily overcome by the other Sphericall Figure.

This is the Principal point in the present Question, in which I perswade my self to be on the right side.

Therefore, beginning to investigate with the examination of exquisite Experiments that really the Figure doth not a jot alter the descent or Ascent of the same Solids, and having already demonstrated that the greater or less Gravity of the Solid in relation to the Gravity of the *Medium* is the cause of Descent or Ascent: when ever we would make proof of that, which about this Effect the diversity of Figure worketh, its necessary to make the Experiment with Matter wherein variety of Gravities hath no place. For making use of Matters which may be different in their Specifical Gravities, and meeting with varieties of effects of Ascending and Descending, we shall alwayes be left unsatisfied whether that diversity derive it self really from the sole Figure, or else from the divers Gravity also. We may remedy this by takeing one only Matter, that is tractable and easily reduceable into every sort of Figure. Moreover, it will be an excellent expedient to take a kinde of Matter, exactly alike in Gravity unto the Water: for that Matter, as far as pertaines to the Gravity, is indifferent either to Ascend or Descend; so that we may presently observe any the least difference that derives it self from the diversity of Figure.

An Experiment in Wax, that proveth Figure to have no Operation in Natation & Submersion.

Now to do this, Wax is most apt, which, besides its incapacity of receiveing any sensible alteration from its imbibing of Water, is ductile or pliant, and the same piece is easily reduceable into all Figures: and being *in Specie* a very inconsiderable matter inferiour in Gravity to the Water, by mixing therewith a little of the fileings of Lead it is reduced to a Gravity exactly equall to that of the Water.

This Matter prepared, and, for example, a Ball being made thereof as bigge as an Orange or biger, and that made so grave as to sink to the bottom, but so lightly, that takeing thence one only Grain of Lead, it returns to the top, and being added, it submergeth to the bottom, let the same Wax afterwards be made into a very broad and thin Flake or Cake; and then, returning to make the same Experiment, you shall see that it being put to the bottom, it shall, with the Grain of Lead rest below, and that Grain deducted, it shall ascend to the very Surface, and added again it shall dive to the bottom. And this same effect shall happen alwaies in all sort of Figures, as wel regular as irregular: nor shall you ever finde any that will swim without the removall of the Grain of Lead, or sinke to the bottom unless it be added: and, in short, about the going or not going to the Bottom, you shall discover no diversity, although, indeed, you shall about the quick and slow descent: for the more expatiated and distended Figures move more slowly aswel in the diveing to the bottom as in the rising to the top; and the other more contracted and compact Figures, more speedily. Now I know not what may be expected from the diversity of Figures, if the most contrary to one another operate not so much as doth a very small Grain of Lead, added or removed.

An objection against the Experiment in Water.

Me thinkes I hear some of the Adversaries to raise a doubt upon my produced Experiment. And first that they offer to my consideration, that the Figure, as a Figure simply, and disjunct from the Matter workes not any effect, but requires to be conjoyned with the Matter; and, furthermore, not with every Matter, but with those only, where-with it may be able to execute the desired operation. Like as we see it verified by Experience, that the Acute and sharp Angle is more apt to cut, than the Obtuse; yet alwaies provided, that both the one and the other, be joyned with a Matter apt to cut, as for example, with Steel. Therefore, a Knife with a fine and sharp edge, cuts Bread or Wood with much ease, which it will not do, if the edge be blunt and thick: but he that will instead of Steel, take Wax, and mould it into a Knife, undoubtedly shall never know the effects of sharp and blunt edges: because neither of them will cut, the Wax being unable by reason of its flexibility, to overcome the hardness of the Wood and Bread. And, therefore, applying the like discourse to our purpose, they say, that the difference of Figure will shew different effects, touching Natation and Submersion, but not conjoyned with any kind of Matter, but only with those Matters which, by their Gravity, are apt to resist the Velocity of the water, whence he that would elect for the Matter, Cork or other light wood, unable, through its Levity, to superate the Crassitude of the water, and of that Matter should forme Solids of divers Figures, would in vain seek to find out what operation Figure hath in Natation or Submersion; because all would swim, and that not through any property of this or that Figure, but through the debility of the Matter, wanting so much Gravity, as is requisite to superate and overcome the Density and Crassitude of the water.

Its needfull, therefore, if wee would see the effect wrought by the Diversity of Figure, first to make choice of a Matter of its nature apt to penetrate the Crassitude of the water. And, for this effect, they have made choice of such a Matter, as fit, that being readily reduced into Sphericall Figure, goes to the Bottom; and it is Ebony, of which they afterwards making a small Board or Splinter, as thin as a Lath, have illustrated how that this, put upon the Surface of the water, rests there without descending to the Bottom: and making, on the otherside, of the same wood a Ball, no less than a hazell Nut, they shew, that this swims not, but descendes. From which Experiment, they think they may frankly conclude, that the Breadth of the Figure in the flat Lath or Board, is the cause of its not descending to the Bottom, for as much as a Ball of the same Matter, not different from the Board in any thing but in Figure, submergeth in the same water to the Bottom. The discourse and the Experiment hath really so much of probability and likelyhood of truth in it, that it would be no wonder, if many perswaded by a certain cursory observation, should yield credit to it; nevertheless, I think I am able to discover, how that it is not free from falacy.

Beginning, therefore, to examine one by one, all the particulars that have been produced, I say, that Figures, as simple Figures, not only operate not in naturall things, but neither are they ever seperated from the Corporeall substance: nor have I ever alledged them stript of sensible Matter, like as also I freely admit, that in our endeavouring to examine the Diversity of Accidents, dependant upon the variety of Figures, it is necessary to apply them to Matters, which obstruct not the various operations of those various Figures: and I admit and grant, that I should do very ill; if I would experiment the influence of Acutenesse of edge with a Knife of Wax, applying it to cut an Oak, because there is no Acuteness in Wax able to cut that very hard wood. But yet such an Experiment of this Knife, would not be besides the purpose, to cut curded Milk, or other very yielding Matter: yea, in such like Matters, the Wax is more commodious than Steel; for finding the diversity depending upon Angles, more or less Acute, for that Milk is indifferently cut with a Raisor, and with a Knife, that hath a blunt edge. It needs, therefore, that regard be had, not only to the hardness, solidity or Gravity of Bodies, which under divers figures, are to divide and penetrate some Matters, but it forceth also, that regard be had, on the other side, to the Resistance of the Matters, to be divided and penetrated. But since I have in making the Experiment concerning our Contest; chosen a Matter which penetrates the Resistance of the water; and in all figures descendes to the Bottome, the Adversaries can charge me with no defect; yea, I have propounded so much a more excellent Method than they, in as much as I have removed all other Causes, of descending or not descending to the Bottom, and retained the only sole and pure variety of Figures, demonstrating that the same Figures all descende with the only alteration of a Grain in weight: which Grain being removed, they return to float and swim; it is not true, therefore, (resuming the Example by them introduced) that I have gon{e} about to experiment the efficacy of Acuteness, in cutting with Matters unable to cut, but with Matters proportioned to our occasion, since they are subjected to no other variety, then that alone which depends on the Figure more or less acute.

The answer to the Objection against the Experiment of the Wax.

But let us proceed a little farther, and observe, how that indeed the Consideration, which, they say, ought to be had about the Election of the Matter, to the end, that it may be proportionate for the making of our experiment, is needlessly introduced, declaring by the example of Cutting, that like as Acuteness is inefficient to cut, unless when it is in a Matter hard and apt to superate the Resistance of the wood or other Matter, which we intend to cut; so the aptitude of descending or not descending in water, ought and can only be known in those Matters, that are able to overcome the Renitence, and superate the Crassitude of the water. Unto which, I say, that to make distinction and election, more of this than of that Matter, on which to impress the Figures for cutting or penetrating this or that Body, as the solidity or obdurateness of the said Bodies shall be greater or less, is very necessary: but withall I subjoyn, that such distinction, election and caution would be superfluous and unprofitable, if the Body to be cut or penetrated, should have no Resistance, or should not at all withstand the Cutting or Penitration: and if the Knife were to be used in cutting a Mist or Smoak, one of Paper would be equally serviceable with one of *Damascus* Steel: and so by reason the water hath not any Resistance against the Penitration of any Solid Body, all choice of Matter is superfluous and needless, and the Election which I said above to have been well made of a Matter reciprocall in Gravity to water, was not because it was necessary, for the overcoming of the crassitude of the water, but its Gravity, with which only it resists the sinking of Solid Bodies: and for what concerneth the Resistance of the crassitude, if we narrowly consider it, we shall find that all Solid Bodies, as well those that sink, as those that swim, are indifferently accomodated and apt to bring us to the knowledge of the truth in question. Nor will I be frighted out of the belief of these Conclusions, by the Experiments which may be produced against me, of many severall Woods, Corks, Galls, and, moreover, of subtle slates and plates of all sorts of Stone and Mettall, apt by means of their Naturall Gravity, to move towards the Centre of the Earth, the which, nevertheless, being impotent, either through the Figure (as the Adversaries thinke) or through Levity, to break and penetrate the Continuity of the parts of the water, and to distract its union, do continue to swimm without submerging in the least: nor on the other side, shall the Authority of *Aristotle* move me, who in more than one place, affirmeth the contrary to this, which Experience shews me.

No Solid of such Levity, nor of such Figure, but that it doth penetrate the Crassitude of the Water.

Bodies of all Figures, laid upon the water, do penetrate its Crassitude, and in what proportion.

I return, therefore, to assert, that there is not any Solid of such Levity, nor of such Figure, that being put upon the water, doth not divide and penetrate its Crassitude: yea if any with a more perspicatious eye, shall return to observe more exactly the thin Boards of Wood, he shall see them to be with part of their thickness under water, and not only with their inferiour Superficies, to kisse the Superiour of the water, as they of necessity must have believed, who have said, that such Boards submerge not, as not being able to divide the Tenacity of the parts of the water: and, moreover, he shall see, that subtle shivers of Ebony, Stone or Metall, when they float, have not only broak the Continuity of the water, but are with all their thickness, under the Surface of it; and more and more, according as the Matters are more grave: so that a thin Plate of Lead, shall be lower than the Surface of the circumfused water, by at least twelve times the thickness of the Plate, and Gold shall dive below the Levell of the water, almost twenty times the thickness of the Plate, as I shall anon declare.

But let us proceed to evince, that the water yields and suffers it self to be penetrated by every the lightest Body; and therewithall demonstrate, how, even by Matters that submerge not, we may come to know that Figure operates nothing about the going or not going to the Bottom, seeing that the water suffers it self to be penetrated equally by every Figure.

The Experiment of a Cone, demitted with its Base, and after with its Point downwards.

Make a Cone, or a Piramis of Cypress, of Firre, or of other Wood of like Gravity, or of pure Wax, and let its height be somewhat great, namely a handfull, or more, and put it into the water with the Base downwards: first, you shall see that it will penetrate the water, nor shall it be at all impeded by the largeness of the Base, nor yet shall it sink all under water, but the part towards the point shall lye above it: by which shall be manifest, first, that that Solid forbeares not to sink out of an inability to divide the Continuity of the water, having already divided it with its broad part, that in the opinion of the Adversaries is the less apt to make the division. The Piramid being thus fixed, note what part of it shall be submerged, and revert it afterwards with the point downwards, and you shall see that it shall not dive into the water more than before, but if you observe how far it shall sink, every person expert in Geometry, may measure, that those parts that remain out of the water, both in the one and in the other Experiment are equall to an hair: whence he may manifestly conclude, that the acute Figure which seemed most apt to part and penetrate the water, doth not part or penetrate it more than the large and spacious.

And he that would have a more easie Experiment, let him take two Cylinders of the same Matter, one long and small, and the other short, but very broad, and let him put them in the water, not distended, but erect and endways: he shall see, if he diligently measure the parts of the one and of the other, that in each of them the part submerged, retains exactly the same proportion to that out of the water, and that no greater part is submerged of that long and small one, than of the other more spacious and broad: howbeit, this rests upon a very large, and that upon a very little Superficies of water: therefore the diversity of Figure, occasioneth neither facility, nor difficulty, in parting and penetrating the Continuity of the water, and, consequently, cannot be the Cause of the Natation or Submersion. He may likewise discover the non operating of variety of Figures, in arising from the Bottom of the water, towards the Surface, by taking Wax, and tempering it with a competent quantity of the filings of Lead, so that it may become a considerable matter graver than the water: then let him make it into a Ball, and thrust it unto the Bottom of the water; and fasten to it as much Cork, or other light matter, as just serveth to raise it, and draw it towards the Surface: for afterwards changing the same Wax into a thin Cake, or into any other Figure, that same Cork shall raise it in the same manner to a hair.

This silenceth not my Antagonists, but they say, that all the discourse hitherto made by me little importeth to them, and that it serves their turn, that they have demonstrated in one only particular, and in what matter, and under what Figure pleaseth them, namely, in a Board and in a Ball of Ebony, that this put in the water, descends to the Bottom, and that stays atop to swim: and the Matter being the same, and the two Bodies differing in nothing but in Figure, they affirm, that they have with all perspicuity demonstrated and sensibly manifested what they undertook; and lastly, that they have obtained their intent. Nevertheless, I believe, and thinke, I can demonstrate, that that same Experiment proveth nothing against my Conclusion.

In Experiments of Natation, the Solid is to be put into, not upon the water.

And first, it is false, that the Ball descends, and the Board not: for the Board shall also descend, if you do to both the Figures, as the words of our Question requireth; that is, if you put them both into the water.

The Question of Natation stated.

The words were these. That the Antagonists having an opinion, that the Figure would alter the Solid Bodies, in relation to the descending or not descending, ascending or not ascending in the same Medium, *as* v. gr. *in the same water, in such sort, that, for Example, a Solid that being of a Sphericall Figure, shall descend to the Bottom, being reduced into some other Figure, shall not descend: I holding the contrary, do affirm, that a Corporeall Solid Body, which reduced into a Sphericall Figure, or any other, shall go to the Bottom, shall do the like under whatsoever other Figure, {&}c.*

Place defined according to Aristotle.

But to be in the water, implies to be placed in the water, and by *Aristotles* own Definition of place, to be placed, importeth to be invironed by the Superficies of the Ambient Body, therefore, then shall the two Figures be in the water, when the Superficies of the water, shall imbrace and inviron them: but when the Adversaries shew the Board of Ebony not descending to the Bottom, they put it not into the water, but upon the water, where being by a certain impediment (as by and by we will shew) retained, it is invironed, part by water, and part by air, which thing is contrary to our agreement, that was, that the Bodies should be in the water, and not part in water, and part in air.

The which is again made manifest, by the questions being put as well about the things which go to the Bottom, as those which arise from the

Bottom to swimme, and who sees not that things placed in the Bottom, must have water about them.

The confutation of the Experiment in the Ebany.

It is now to be noted, that the Board of Ebany and the Ball, put into the water, both sink, but the Ball more swiftly, and the Board more slowly; and slower and slower, according as it shall be more broad and thin, and of this Tardity the breadth of the Figure is the true Cause: But these broad Boards that slowly descend, are the same, that being put lightly upon the water, do swimm: Therefore, if that were true which the Adversaries affirm, the same numerical Figure, would in the same numericall water, cause one while Rest, and another while Tardity of Motion, which is impossible: for every perticular Figure which descends to the Bottom, hath of necessity its own determinate Tardity and slowness, proper and naturall unto it, according to which it moveth, so that every other Tardity, greater or lesser is improper to its nature: if, therefore, a Board, as suppose of a foot square, descendeth naturally with six degrees of Tardity, it is impossible, that it should descend with ten or twenty, unless some new impediment do arrest it. Much less can it, by reason of the same Figure rest, and wholly cease to move; but it is necessary, that when ever it resteth, there do some greater impediment intervene than the breadth of the Figure. Therefore, it must be somewhat else, and not the Figure, that stayeth the Board of Ebany above water, of which Figure the only Effect is the retardment of the Motion, according to which it descendeth more slowly than the Ball. Let it be confessed, therefore, rationally discoursing, that the true and sole Cause of the Ebanys going to the Bottom, is the excess of its Gravity above the Gravity of the water: and the Cause of the greater or less Tardity, the breadth of this Figure, or the contractedness of that: but of its Rest, it can by no means be allowed, that the quallity of the Figure, is the Cause thereof: aswell, because, making the Tardity greater, according as the Figure more dilateth, there cannot be so immense a Dilatation, to which there may not be found a correspondent immence Tardity without redusing it to Nullity of Motion; as, because the Figures produced by the Antagonists for effecters of Rest, are the self same that do also go to the Bottom.

I will not omit another reason, founded also upon Experience, and if I deceive not my self, manifestly concluding, how that the Introduction of the breadth or amplitude of Figure, and the Resistance of the water against penetration, have nothing to do in the Effect of descending, or ascending, or resting in the water.[4] Take a piece of wood or other Matter, of which a Ball ascends from the Bottom of the water to the Surface, more slowly than a Ball of Ebony of the same bignesse, so that it is manifest, that the Ball of Ebony more readily divideth the water in descending, than the other in ascending; as for Example, let the Wood be Walnut-tree. Then take a Board of Walnut-tree, like and equall to that of Ebony of the Antagonists, which swims; and if it be true, that this floats above water, by reason of the Figure, unable through its breadth, to pierce the Crassitude of the same, the other of Wallnut-tree, without all question, being thrust unto the Bottom, will stay there, as less apt, through the same impediment of Figure, to divide the said Resistance of the water. But if we shall find, and by experience see, that not only the thin Board, but every other Figure of the same Wallnut-tree will return to float, as undoubtedly we shall, then I must desier my opposers to forbear to attribute the floating of the Ebony, unto the Figure of the Board, in regard that the Resistance of the water is the same, as well to the ascent, as to the descent, and the force of the Wallnut-trees ascension, is lesse than the Ebonys force in going to the Bottom.

An Experiment in Gold, to prove the non-operating of Figure in Natation and Submersion.

Nay, I will say more, that if we shall consider Gold in comparison of water, we shall find, that it exceeds it in Gravity almost twenty times, so that the Force and Impetus, wherewith a Ball of Gold goes to the Bottom, is very great. On the contrary, there want not matters, as Virgins Wax, and some Woods, which are not above a fiftieth part less grave than water, whereupon their Ascension therein is very slow, and a thousand times weaker than the *Impetus* of the Golds descent: yet notwithstanding, a plate of Gold swims without descending to the Bottom, and, on the contrary, we cannot make a Cake of Wax, or thin Board of Wood, which put in the Bottom of the Water, shall rest there without ascending. Now if the Figure can obstruct the Penetration, and impede the descent of Gold, that hath so great an *Impetus*, how can it choose but suffice to resist the same Penetration of the other matter in

---

[4] The Figure & Resistance of the Medium against Division, have nothing to do with the Effect of Natation or Submersion, by an Experiment in Wallnut tree.

ascending, when as it hath scarce a thousandth part of the *Impetus* that the Gold hath in descending? Its therefore, necessary, that that which suspends the thin Plate of Gold, or Board of Ebony, upon the water, be some thing that is wanting to the other Cakes and Boards of Matters less grave than the water; since that being put to the Bottom, and left at liberty, they rise up to the Surface, without any obstruction: But they want not for flatness and breadth of Figure: Therefore, the *spaciousnesse of the Figure, is not that which makes the Gold and Ebony to swim.

And, because, that the excess of their Gravity above the Gravity of the water, is questionless the Cause of the sinking of the flat piece of Ebony, and the thin Plate of Gold, when they go to the Bottom, therefore, of necessity, when they float, the Cause of their staying above water, proceeds from Levity, which in that case, by some Accident, peradventure not hitherto observed, cometh to meet with the said Board, rendering it no longer as it was before, whilst it did sink more ponderous than the water, but less.

Now, let us return to take the thin Plate of Gold, or of Silver, or the thin Board of Ebony, and let us lay it lightly upon the water, so that it stay there without sinking, and diligently observe its effect. And first, see how false the assertion of *Aristotle*, and our oponents is, to wit, that it stayeth above water, through its unability to pierce and penetrate the Resistance of the waters Crassitude: for it will manifestly appear, not only that the said Plates have penetrated the water, but also that they are a considerable matter lower than the Surface of the same, the which continueth eminent, and maketh as it were a Rampert on all sides, round about the said Plates, the profundity of which they stay swimming: and, according as the said Plates shall be more grave than the water, two, four, ten or twenty times, it is necessary, that their Superficies do stay below the universall Surface of the water, so much more, than the thickness of those Plates, as we shal more distinctly shew anon. In the mean space, for the more easie understanding of

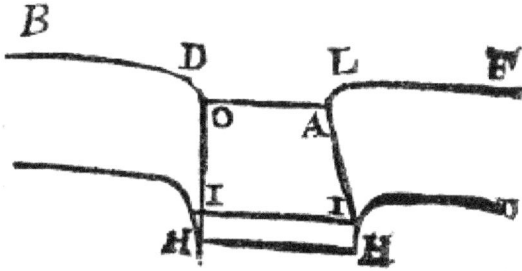

what   I say, observe with me a little the present Scheme: in which let us suppose the Surface of the water to be distended, according to the Lines F L D B, upon which if one shall put a board of matter specifically more grave than water, but so lightly that it submerge not, it shall not rest any thing above, but shall enter with its whole thickness into the water: and, moreover, shall sink also, as we see by the Board A I, O I, whose breadth is wholly sunk into the water, the little Ramperts of water L A and D O incompassing it, whose Superficies is notably higher than the Superficies of the Board. See now whether it be true, that the said Board goes not to the Bottom, as being of Figure unapt to penetrate the Crassitude of the water.

Why solids having penitrated the Water, do not proceed to a totall Submersion.

But, if it hath already penetrated, and overcome the Continuity of the water, & is of its own nature more grave than the said water, why doth it not proceed in its sinking, but stop and suspend its self within that little dimple or cavitie, which with its ponderosity it hath made in the water? I answer; because that in submerging it self, so far as till its Superficies come to the Levell with that of the water, it loseth a part of its Gravity, and loseth the rest of it as it submergeth & descends beneath the Surface of the water, which maketh Ramperts and Banks round about it, and it sustaines this loss by means of its drawing after it, and carrying along with it, the Air that is above it, and by Contact adherent to it, which Air succeeds to fill the Cavity that is invironed by the Ramperts of water; so that that which in this case descends and is placed in the water, is not only the Board of Ebony or Plate of Iron, but a composition of Ebony and Air, from which resulteth a Solid no longer superiour in Gravity to the water, as was the simple Ebony, or the simple Gold. And, if we exactly consider, what, and how great the Solid is, that in this Experiment enters into the water, and contrasts with the Gravity of the same, it will be found to be all that which we find to be beneath the Surface of the water, the which is an aggregate and Compound of a Board of Ebony, and of almost the like quantity of Air, or a Mass compounded of a Plate of Lead, and ten or twelve times as much Air. But, Gentlemen, you that are my Antagonists in our Question, we require the Identity of Matter, and the alteration only of the Figure; therefore, you must remove that Air, which being conjoyned with the Board, makes it become another Body less grave than the Water, and put only the Ebony into the Water, and you shall certainly see the Board descend to the Bottom; and, if that do not happen, you have got the day. And to seperate the Air from the Ebony, there needs no more but only to bath the Superficies of the said Board with the same Water: for the Water being thus interposed between the Board and the Air, the other circumfused Water shall run together without any impediment, and shall receive into it the sole and bare Ebony, as it was to do.

But, me thinks I hear some of the Adversaries cunningly opposing this, and telling me, that they will not yield, by any means, that their Board be wetted, because the weight added thereto by the Water, by making it heavier than it was before, draws it to the Bottom, and that the addition of new weight is contrary to our agreement, which was, that the Matter be the same.

To this, I answer, first; that treating of the operation of Figure in Bodies put into the Water, none can suppose them to be put into the Water without being wet; nor do I desire more to be done to the Board, then I will give you leave to do to the Ball. Moreover, it is untrue, that the Board sinks by vertue of the new Weight added to it by the Water, in the single and slight bathing of it: for I will put ten or twenty drops of Water upon the same Board, whilst it is sustained upon the water; which drops, because not conjoyned with the other Water circumfused, shall not so encrease the weight of it, as to make it sink: but if the Board being taken out, and all the water wiped off that was added thereto, I should bath all its Superficies with one only very small drop, and put it again upon the water, without doubt it shall sink, the other Water running to cover it, not being retained by the superiour Air; which Air by the interposition of the thin vail of water, that takes away its Contiguity unto the Ebony, shall without Renitence be seperated, nor doth it in the least oppose the succession of the other Water: but rather, to speak better, it shall descend freely; because it shall be all invironed and covered with water, as soon as its superiour Superficies, before vailed with water, doth arrive to the Levell of the universall Surface of the said water. To say, in the next place, that water can encrease the weight of things that are demitted into it, is most false; for water hath no Gravity in water, since it descends not: yea, if we would well consider what any immense Mass of water doth put upon a grave Body; that is placed in it, we shall find experimentally, that it, on the contrary, will rather in a great part deminish the weight of it, and that we may be able to lift an huge Stone from the Bottom of the water, which the water being removed, we are not able to stir. Nor let them tell me by way of reply, that although the superposed water augment not the Gravity of things that are in it, yet it increaseth the ponderosity of those that swim, and are part in the water and part in the Air, as is seen, for Example, in a Brass Ketle, which whilst it is empty of water, and replenished only with Air shall swim, but pouring of Water therein, it shall become so grave, that it shall sink to the Bottom, and that by reason of the new weight added thereto. To this I will return answer, as above, that the Gravity of the Water, contained in the Vessel is not that which sinks it to the Bottom, but the proper Gravity of the Brass, superiour to the Specificall Gravity of the Water: for if the Vessel were less grave than water, the Ocean would not suffice to submerge it. And, give me leave to repeat it again, as the fundamentall and principall point in this Case, that the Air contained in this Vessel before the infusion of the Water, was that which kept it a-float, since that there was made of it, and of the Brass, a Composition less grave than an equall quantity of Water: and the place that the Vessel occupyeth in the Water whilst it floats, is not equall to the Brass alone, but to the Brass and to the Air together, which filleth that part of the Vessel that is below the Levell of the water: Moreover, when the Water is infused, the Air is removed, and there is a composition made of Brass and of water, more grave *in specie* than the simple water, but not by vertue of the water infused, as having greater Specifick Gravity than the other water, but through the proper Gravity of the Brass, and through the alienation of the Air. Now, as he that should say that Brass, that by its nature goes to the Bottom, being formed into the Figure of a Ketle, acquireth from that Figure a vertue of lying in the Water without sinking, would say that which is false; because that Brass fashioned into any whatever Figure, goeth always to the Bottom, provided, that that which is put into the water be simple Brass; and it is not the Figure of the Vessel that makes the Brass to float, but it is because that that is not purely Brass which is put into the water, but an aggregate of Brass and of Air: so is it neither more nor less false, that a thin Plate of Brass or of Ebony, swims by vertue of its dilated & broad Figure: for the truth is, that it bares up without submerging, because that that which is put in the water, is not pure Brass or simple Ebony, but an aggregate of Brass and Air, or of Ebony and Air. And, this is not contrary unto my Conclusion, the which, (having many a time seen Vessels of Mettall, and thin pieces of diverse grave Matters float, by vertue of the Air conjoyned with them) did affirm, That Figure was not the Cause of the Natation or Submersion of such Solids as were placed in the water. Nay more, I cannot omit, but must tell my Antagonists, that this new conceit of denying that the Superficies of the Board should be bathed, may beget in a third person an opinion of a poverty of Arguments of defence on their part, since that such bathing was never insisted upon by them in the beginning of our Dispute, and was not questioned in the least, being that the Originall of the discourse arose upon the swiming of Flakes of Ice, wherein it would be simplicity to require that their Superficies might bedry: besides, that whether these pieces of Ice be wet or dry they alwayes swim, and as the Adversaries say, by reason of the Figure.

Some peradventure, by way of defence, may say, that wetting the Board of Ebony, and that in the superiour Superficies, it would, though of it self unable to pierce and penetrate the water, be born downwards, if not by the weight of the additionall water, at least by that desire and propension that the superiour parts of the water have to re-unite and rejoyn themselves: by the Motion of which parts, the said Board cometh in a certain manner, to be depressed downwards.

The Bathed Solid descends not out of any affectation of union in the upper parts of the water.

This weak Refuge will be removed, if we do but consider, that the repugnancy of the inferiour parts of the water, is as great against Disunion, as the Inclination of its superiour parts is to union: nor can the uper unite themselves without depressing the board, nor can it descend without disuniting the parts of the nether Water: so that it doth follow, by necessary consequence, that for those respects, it shall not descend. Moreover, the same that may be said of the upper parts of the water, may with equall reason be said of the nether, namely, that desiring to unite, they shall force the said Board upwards.

Happily, some of these Gentlemen that dissent from me, will wonder, that I affirm, that the contiguous superiour Air is able to sustain that Plate of Brass or of Silver, that stayeth above water; as if I would in a certain sence allow the Air, a kind of Magnetick vertue of sustaining the grave Bodies, with which it is contiguous. To satisfie all I may, to all doubts, I have been considering how by some other sensible Experiment I might demonstrate, how truly that little contiguous and superiour Air sustaines those Solids, which being by nature apt to descend to the Bottom, being placed lightly on the water submerge not, unless they be first thorowly bathed; and have found, that one of these Bodies having descended to the Bottom, by conveighing to it (without touching it in the least) a little Air, which conjoyneth with the top of the same, it becometh sufficient, not only, as before to sustain it, but also to raise it, and to carry it back to the top, where it stays and abideth in the same manner, till such time, as the assistance of the conjoyned Air is taken away. And to this effect, I have taken a Ball of Wax, and made it with a little Lead, so grave, that it leasurely descends to the Bottom, making with all its Superficies very smooth and pollite: and this being put gently into the water, almost wholly submergeth, there remaining vissible only a little of the very top, the which so long as it is conjoyned with the Air, shall retain the Ball atop, but the Contiguity of the Air taken away by wetting it, it shall descend to the Bottom and there remain. Now to make it by vertue of the Air, that before sustained it to return again to the top, and stay there, thrust into the water a Glass reversed with the mouth downwards, the which shall carry with it the Air it contains, and move this towards the Ball, abasing it till such time that you see, by the transparency of the Glass, that the contained Air do arrive to the summity of the *B*all: then gently withdraw the Glass upwards, and you shall see the *B*all to rise, and afterwards stay on the top of the water, if you carefully part the Glass and the water without overmuch commoving and disturbing it. There is, therefore, a certain affinity between the Air and other Bodies, which holds them unied, so, that they seperate not without a kind of violence. The same likewise is seen in the water; for if we shall wholly submerge some Body in it, so that it be thorowly bathed, in the drawing of it afterwards gently out again, we shall see the water follow it, and rise notably above its Surface, before it seperates from it. Solid Bodies, also, if they be equall and alike in Superficies, so, that they make an exact Contact without the interposition of the least Air, that may part them in the seperation and yield untill that the ambient *Medium* succeeds to replenish the place, do hold very firmly conjoyned, and are not to be seperated without great force but, because, the Air, Water, and other Liquids, very expeditiously shape themselves to contact with any Solid *Bodies*, so that their Superficies do exquisitely adopt themselves to that of the Solids, without any thing remaining between them, therefore, the effect of this Conjunction and Adherence is more manifestly and frequently observed in them, than in hard and inflexible Bodies, whose Superficies do very rarely conjoyn with exactness of Contact. This is therefore that Magnetick vertue, which with firm Connection conjoyneth all Bodies, that do touch without the interposition of flexible fluids; and, who knows, but that that a Contact, when it is very exact, may be a sufficient Cause of the Union and Continuity of the parts of a naturall *Body*?

Now, pursuing my purpose, I say; that it needs not, that we have recourse to the Tenacity, that the parts of the water have amongst themselves, by which they resist and oppose Division, Distraction, and Seperation, because there is no such Coherence and Resistance of Division for if there were, it would be no less in the internall parts than in those nearer the superiour or externall Surface, so that the same Board, finding alwayes the same Resistance and Renitence, would no less stop in the middle of the water than about the Surface, which is false. Moreover, what Resistance can we place in the Continuity of the water, if we see that it is impossible to find any Body of whatsoever Matter, Figure or Magnitude, which being put into the water, shall be obstructed and impeded by the Tenacity of the parts of the water to one another, so, but that it is moved upwards or downwards, according as the Cause of their Motion transports it? And, what greater proof of it can we desier, than that which we daily see in Muddy waters, which being put into Vessels to be drunk, and being, after some hours setling, still, as we say, thick in the end, after four or six dayes they are wholly setled, and become pure and clear? Nor can their Resistance of Penetration stay those impalpable and insensible Atomes of Sand, which by reason of their exceeding small force, spend six dayes in descending the space of half a yard.

Nor let them say, that the seeing of such small Bodies, consume six dayes in descending so little a way, is a sufficient Argument of the Waters Resistance of Division; because that is no resisting of Division, but a retarding of Motion; and it would be simplicity to say, that a thing opposeth Division, and that in the same instant, it permits it self to be divided: nor doth the Retardation of Motion at all favour the Ad-

versaries cause, for that they are to instance in a thing that wholly prohibiteth Motion, and procureth Rest; it is necessary, therefore, to find out Bodies that stay in the water, if one would shew its repugnancy to Division, and not such as move in it, howbeit but slowly.

What then is this Crassitude of the water, with which it resisteth Division? What, I beseech you, should it be, if we (as we have said above) with all diligence attempting the reduction of a Matter into so like a Gravity with the water, that forming it into a dilated Plate it rests suspended as we have said, between the two waters, it be impossible to effect it, though we bring them to such an Equiponderance, that as much Lead as the fourth part of a Grain of Musterd-seed, added to the same expanded Plate, that in Air [*i. e. out of the water*] shall weigh four or six pounds, sinketh it to the Bottom, and being substracted, it ascends to the Surface of the water? I cannot see, (if what I say be true, as it is most certain) what minute vertue and force we can possibly find or imagine, to which the Resistance of the water against Division and Penetration is not inferiour; whereupon, we must of necessity conclude that it is nothing: because, if it were of any sensible power, some large Plate might be found or compounded of a Matter alike in Gravity to the water, which not only would stay between the two waters; but, moreover, should not be able to descend or ascend without notable force. We may likewise collect the same from an other Experiment, shewing that the Water gives way also in the same manner to transversall Division; for if in a setled and standing water we should place any great Mass that goeth not to the bottom, drawing it with a single Womans Hair, we might carry it from place to place without any opposition, and this whatever Figure it hath, though that it possess a great space of water, as for instance, a great Beam would do moved sideways. Perhaps some might oppose me and say, that if the Resistance of water against Division, as I affirm, were nothing; Ships should not need such a force of Oars and Sayles for the moving of them from place to place in a tranquile Sea, or standing Lake. To him that should make such an objection, I would reply, that the water contrasteth not against, nor simply resisteth Division, but a sudden Division, and with so much greater Renitence, by how much greater the Velocity is: and the Cause of this Resistance depends not on Crassitude, or any other thing that absolutely opposeth Division, but because that the parts of the water divided, in giving way to that Solid that is moved in it, are themselves also necessitated locally to move, some to the one side, and some to the other, and some downwards: and this must no less be done by the waves before the Ship, or other Body swimming through the water, than by the posteriour and subsequent; because, the Ship proceeding forwards, to make it self a way to receive its Bulk, it is requisite, that with the Prow it repulse the adjacent parts of the water, as well on one hand as on the other, and that it move them as much transversly, as is the half of the breadth of the Hull: and the like removall must those waves make, that succeeding the Poump do run from the remoter parts of the Ship towards those of the middle, successively to replenish the places, which the Ship in advancing forwards, goeth, leaving vacant. Now, because, all Motitions are made in Time, and the longer in greater time: and it being moreover true, that those Bodies that in a certain time are moved by a certain power such a certain space, shall not be moved the same space, and in a shorter Time, unless by a greater Power: therefore, the broader Ships move slower than the narrower, being put on by an equall Force: and the same Vessel requires so much greater force of Wind, or Oars, the faster it is to move.

But yet for all this, any great Mass swimming in a standing Lake, may be moved by any petit force; only it is true, that a lesser force more slowly moves it: but if the waters Resistance of Division, were in any manner sensible, it would follow, that the said Mass, should, notwithstanding the percussion of some sensible force, continue immoveable, which is not so. Yea, I will say farther, that should we retire our selves into the more internall contemplation of the Nature of water and other Fluids, perhaps we should discover the Constitution of their parts to be such, that they not only do not oppose Division, but that they have not any thing in them to be divided: so that the Resistance that is observed in moving through the water, is like to that which we meet with in passing through a great Throng of People, wherein we find impediment, and not by any difficulty in the Division, for that none of those persons are divided whereof the Croud is composed, but only in moving of those persons side-ways which were before divided and disjoyned: and thus we find Resistance in thrusting a Stick into an heap of Sand, not because any part of the Sand is to be cut in pieces, but only to be moved and raised. Two manners of Penetration, therefore, offer themselves to us, one in Bodies, whose parts were continuall, and here Division seemeth necessary, the other in the aggregates of parts not continuall, but contiguous only, and here there is no necessity of dividing but of moving only. Now, I am not well resolved, whether water and other Fluids may be esteemed to be of parts continuall or contiguous only; yet I find my self indeed inclined to think that they are rather contiguous (if there be in Nature no other manner of aggregating, than by the union, or by the touching of the extreams:) and I am induced thereto by the great difference that I see between the Conjunction of the parts of an hard or Solid Body, and the Conjunction of the same parts when the same Body shall be made Liquid and Fluid: for if, for example, I take a Mass of Silver or other Solid and hard Mettall, I shall in dividing it into two parts, find not only the resistance that is found in the moving of it only, but an other incomparably greater, dependent on that vertue, whatever it be, which holds the parts united: and so if we would divide again those two parts into other two, and successively into others and others, we should still find a like Resistance, but ever less by how much smaller the parts to be divided shall be; but if, lastly, employing most subtile and acute Instruments, such as are the most tenuous parts of the Fire, we shall resolve it (perhaps) into its last and least Particles, there shall not be left in them any longer either Resistance of Division, or so much as a capacity of being farther divided, especially by Instruments more grosse than the acuities of Fire: and what Knife or Rasor put into well melted Silver can we finde, that will divide a thing which surpasseth the separating power of Fire? Certainly none: because either the whole shall be reduced to the most minute and ultimate Divisions, or if there remain parts capable still of other Subdidivisions, they cannot receive them, but only from acuter Divisors than Fire; but a Stick or Rod of Iron, moved in the melted Metall, is not such a one. Of a like Constitution and Consistence, I account the parts of Water, and other Liquids to be, namely, incapable of Division by reason of their Ienuity; or if not absolutely indivisible, yet at least not to be divided by a Board, or other Solid Body, palpable unto the hand, the Sector being always required to be more sharp than the Solid to be cut. Solid Bodies, therefore, do only move, and not divide the Water, when put into it; whose parts being before divided to the extreamest minuity, and therefore capable of being moved, either many of them at once, or few, or very few, they soon give place to every small Corpuscle, that descends in the same: for that, it being little and light, descending in the Air, and arriving to the Surface of the Water, it meets with Particles of Water more small, and of less Resistance against Motion and Extrusion, than is its own prement and extrusive force; whereupon it submergeth, and moveth such a portion of them, as is proportionate to its Power. There is not, therefore, any Resistance in Water against Division, nay, there is not in it any divisible parts. I adde; moreover, that in case yet there should be any small Resistance found (which is absolutely false) haply in attempting with an Hair to move a very great natant Machine, or in essaying by the addition of one small Grain of Lead to sink, or by removall of it to raise a very broad Plate of Matter, equall in Gravity with Water, (which likewise will not happen, in case we proceed with dexterity) we may observe that that Resistance is a very different thing from that which the Adversaries produce for the Cause of the Natation of the Plate of Lead or Board of Ebony, for that one may make a Board of Ebony, which being put upon the Water swimmeth, and cannot be submerged, no not by the addition of an hundred Grains of Lead put upon the same, and afterwards being bathed, not only sinks, though the said Lead be taken away, but though moreover a quantity of Cork, or of some other light Body fastened to it, sufficeth not to hinder it from sinking unto the bottome: so that you see, that although it were granted that there is a certain small Resistance of Division found in the substance of the Water, yet this hath nothing to do with that Cause which supports the Board above the Water, with a Resistance an hundred times greater than that which men can find in the parts of the Water: nor let them tell me, that only the Surface of the Water hath such Resistance, and not the internall parts, or that such Resistance is found greatest in the beginning of the Submersion, as it also seems that in the beginning, Motion meets with greater opposition, than in the continuance of it; because, first, I will permit, that the Water be stirred, and that the superiour parts be mingled with the middle, and inferiour parts, or that those above be wholly removed, and those in the middle only made use off, and yet you shall see the effect for all that, to be still the same: Moreover, that Hair which draws a Beam through the Water, is likewise to divide the upperparts, and is also to begin the Motion, and yet it begins it, and yet it divides it: and finally, let the Board of Ebony be put in the midway, betwixt the bottome and the top of the Water, and let it there for awhile be suspended and setled, and afterwards let it be left at liberty, and it will instantly begin its Motion, and will continue it unto the bottome. Nay, more, the Board so soon as it is dimitted upon the Water, hath not only begun to move and divide it, but is for a good space dimerged into it.

Let us receive it, therefore, for a true and undoubted Conclusion, That the Water hath not any Renitence against simple Division, and that it is not possible to find any Solid Body, be it of what Figure it will, which being put into the Water, its Motion upwards or downwards, according as it exceedeth, or shall be exceeded by the Water in Gravity (although such excesse and difference be insensible) shall be prohibited, and taken away, by the Crassitude of the said Water. When, therefore, we see the Board of Ebony, or of other Matter, more grave than the Water, to stay in the Confines of the Water and Air, without submerging, we must have recourse to some other Originall, for the investing the Cause of that Effect, than to the breadth of the Figure, unable to over-come the Renitence with which the Water opposeth Division, since there is no Resistance; and from that which is not in being, we can ex-pect no Action. It remains most true, therefore, as we have said before, that this so succeds, for that that which in such manner put upon the water, not the same Body with that which is put *into* the Water: be-cause this which is put *into* the Water, is the pure Board of Ebony, which for that it is more grave than the Water, sinketh, and that which is put *upon* the Water, is a Composition of Ebony, and of so much Air, that both together are specifically less grave than the Water, and there-fore they do not descend.

I will farther confirm this which I say. Gentlemen, my Antagonists, we are agreed, that the excess or defect of the Gravity of the Solid, unto the Gravity of the Water, is the true and proper Cause of Natation or Submersion.

Great Caution to be had in experimenting the operation of Figure in Natation.

Now, if you will shew that besides the former Cause, there is another which is so powerfull, that it can hinder and remove the Submersion of those very Solids, that by their Gravity sink, and if you will say, that this is the breadth or ampleness of Figure, you are oblieged, when ever you would shew such an Experiment, first to make the circumstances certain, that that Solid which you put into the Water, be not less grave *in specie* than it, for if you should not do so, any one might with rea-son say, that not the Figure, but the Levity was the cause of that Natation. But I say, that when you shall dimit a Board of Ebony into the Water, you do not put therein a Solid more grave *in specie* than the Water, but one lighter, for besides the Ebony, there is in the Water a Mass of Air, united with the Ebony, and such, and so light, that of both there results a Composition less grave than the Water: See, therefore, that you remove the Air, and put the Ebony alone into the Water, for so you shall immerge a Solid more grave then the Water, and if this shall not go to the Bottom, you have well Philosophized and I ill.

Now, since we have found the true Cause of the Natation of those Bodies, which otherwise, as being graver than the Water, would descend to the bottom, I think, that for the perfect and distinct knowledge of this business, it would be good to proceed in a way of discovering demonstratively those particular Accidents that do attend these effects, and,

PROBL. I.

To finde what proportion severall Figures of different Matters ought to have, unto the Gravity of the Water, that so they may be able by vertue of the Contiguous Air to stay afloat.

Let, therefore, for better illustration, D F N E be a Vessell, wherein the water is contained, and suppose a Plate or Board, whose thickness is comprehended between the Lines I C and O S, and let it be of Matter exceeding the water in Gravity, so that being put upon the water, it dimergeth and abaseth below the Levell of the said water, leaving the little Banks A I and B C, which are at the greatest height they can be, so that if the Plate I S should but descend any little space farther, the little   Banks   or   Ramparts   would   no   longer   consist,

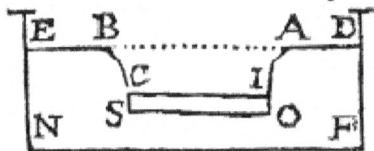

but expulsing the Air A I C B, they would diffuse themselves over the Superficies I C, and would submerge the Plate. The height A I B C is therefore the greatest profundity that the little Banks of water admit of. Now I say, that from this, and from the proportion in Gravity, that the Matter of the Plate hath to the water, we may easily finde of what thickness, at most, we may make the said Plates, to the end, they may be able to bear up above water: for if the Matter of the Plate or Board I S were, for Example, as heavy again as the water, a Board of that Matter shall be, at the most of a thickness equall to the greatest height of the Banks, that is, as thick as A I is high: which we will thus demonstrate. Let the Solid I S be double in Gravity to the water, and let it be a regular Prisme, or Cylinder, to wit, that hath its two flat Superficies, superiour

53

and inferiour, alike and equall, and at Right Angles with the other lat-
erall Superficies, and let its thickness I O be equall to the greatest

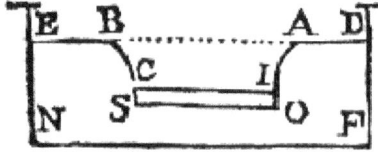

Altitude of the Banks   of water: I say,
that if it be put upon the water, it will not submerge: for the Altitude A
I being equall to the Altitude I O, the Mass of the Air A B C I shall be
equall to the Mass of the Solid C I O S: and the whole Mass A O S B
double to the Mass I S; And since the Mass of the Air A C, neither
encreaseth nor diminisheth the Gravity of the Mass I S, and the Solid I
S was supposed double in Gravity to the water; Therefore as much
water as the Mass submerged A O S B, compounded of the Air A I C
B, and of the Solid I O S C, weighs just as much as the same submerged Mass A O S B: but when such a Mass of water, as is the
submerged part of the Solid, weighs as much as the said Solid, it descends not farther, but resteth, as by (a) *Archimedes*, and above by us,
hath been demonstrated: Therefore, I S shall descend no farther, but
shall rest. And if the Solid I S shall be Sesquialter in Gravity to the
water, it shall float, as long as its thickness be not above twice as much
as the greatest Altitude of the Ramparts of water, that is, of A I. For I S
being Sesquialter in Gravity to the water, and the Altitude O I, being
double to I A, the Solid submerged A O S B, shall be also Sesquialter
in Mass to the Solid I S. And because the Air A C, neither increaseth
nor diminisheth the ponderosity of the Solid I S: Therefore, as much
water in quantity as the submerged Mass A O S B, weighs as much as
the said Mass submerged: And, therefore, that Mass shall rest. And
briefly in generall.

THEOREME. VI.

The proportion of the greatest thickness of Solids, beyond which encreased they sink.

*When ever the excess of the Gravity of the Solid above the Gravity of
the Water, shall have the same proportion to the Gravity of the Water,
that the Altitude of the Rampart, hath to the thickness of the Solid, that
Solid shall not sink, but being never so little thicker it shall.*

Let the Solid I S be superior in Gravity to the water, and of such thickness, that the Altitude of the Rampart A I, be in proportion to the thickness of the Solid I O, as the excess of the Gravity of the said Solid I S, above the Gravity of a Mass of water equall to the Mass I S, is to the Gravity of the Mass of water equall to the Mass I

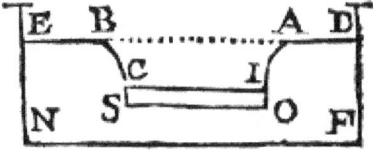

S. I say, that the Solid I S shall not sinke, but being never so little thicker it shall go to the bottom: For being that as A I is to I O, so is the Excess of the Gravity of the Solid I S, above the Gravity of a Mass of water equall to the Mass I S, to the Gravity of the said Mass of water: Therefore, compounding, as A O is to O I, so shall the Gravity of the Solid I S, be to the Gravity of a Mass of water equall to the Mass I S: And, converting, as I O is to O A, so shall the Gravity of a Mass of water equall to the Mass I S, be to the Gravity of the Solid I S: But as I O is to O A, so is a Mass of water I S, to a Mass of water equall to the Mass A B S O: and so is the Gravity of a Mass of water I S, to the Gravity of a Mass of water A S: Therefore as the Gravity of a Mass of water, equall to the Mass I S, is to the Gravity of the Solid I S, so is the same Gravity of a Mass of water I S, to the Gravity of a Mass of Water A S: Therefore the Gravity of the Solid I S, is equall to the Gravity of a Mass of water equall to the Mass A S: But the Gravity of the Solid I S, is the same with the Gravity of the Solid A S, compounded of the Solid I S, and of the Air A B C I. Therefore the whole compounded Solid A O S B, weighs as much as the water that would be comprised in the place of the said Compound A O S B: And, therefore, it shall make an *Equilibrium* and rest, and that same Solid I O S C shall sinke no farther. But if its thickness I O should be increased, it would be necessary also to encrease the Altitude of the Rampart A I, to maintain the due proportion: But by what hath been supposed, the Altitude of the Rampart A I, is the greatest that the Nature of the Water and Air do admit, without the waters repulsing the Air adherent to the Superficies of the Solid I C, and possessing the space A I C B: Therefore, a Solid of greater thickness than I O, and of the same Matter with the Solid I S, shall not rest without submerging, but shall descend to the bottome: which was to be demonstrated. In consequence of this that hath been demonstrated, sundry and various Conclusions may be gathered, by which the truth of my principall Proposition comes to be more and more confirmed, and the imperfection of all former Argumentations touching the present Question cometh to be discovered.

And first we gather from the things demonstrated, that,

THEOREME VII.

The heaviest Bodies may swimme.

All Matters, how heavy soever, even to Gold it self, the heaviest of all Bodies, known by us, may float upon the Water.

Because its Gravity being considered to be almost twenty times greater than that of the water, and, moreover, the greatest Altitude that the Rampart of water can be extended to, without breaking the Contiguity of the Air, adherent to the Surface of the Solid, that is put upon the water being predetermined, if we should make a Plate of Gold so thin, that it exceeds not the nineteenth part of the Altitude of the said Rampart, this put lightly upon the water shall rest, without going to the bottom: and if Ebony shall chance to be in sesquiseptimall proportion more grave than the water, the greatest thickness that can be allowed to a Board of Ebony, so that it may be able to stay above water without sinking, would be seaven times more than the height of the Rampart Tinn, *v. gr.* eight times more grave than water, shall swimm as oft as the thickness of its Plate, exceeds not the 7th part of the Altitude of the Rampart.

He elsewhere cites this as a Proposition, therefore I make it of that number.

And here I will not omit to note, as a second Corrollary dependent upon the things demonstrated, that,

THEOREME VIII.

Natation and Submersion, collected from the thickness, excluding the length and breadth of Plates.

The Expansion of Figure not only is not the Cause of the Natation of those grave Bodies, which otherwise do submerge, but also the deter-

mining what be those Boards of Ebony, or Plates of Iron or Gold that will swimme, depends not on it, rather that same determination is to be collected from the only thickness of those Figures of Ebony or Gold, wholly excluding the consideration of length and breadth, as having no wayes any share in this Effect.

It hath already been manifested, that the only cause of the Natation of the said Plates, is the reduction of them to be less grave than the water, by means of the connexion of that Air, which descendeth together with them, and possesseth place in the water; which place so occupied, if before the circumfused water diffuseth it self to fill it, it be capable of as much water, as shall weigh equall with the Plate, the Plate shall remain suspended, and sinke no farther.

Now let us see on which of these three dimensions of the Solid depends the terminating, what and how much the Mass of that ought to be, that so the assistance of the Air contiguous unto it, may suffice to render it specifically less grave than the water, whereupon it may rest without Submersion. It shall undoubtedly be found, that the length and breadth have not any thing to do in the said determination, but only the height, or if you will the thickness: for, if we take a Plate or Board, as for Example, of Ebony, whose Altitude hath unto the greatest possible Altitude of the Rampart, the proportion above declared, for which cause it swims indeed, but yet not if we never so little increase its thickness; I say, that retaining its thickness, and encreasing its Superficies to twice, four times, or ten times its bigness, or dminishing it by dividing it into four, or six, or twenty, or a hundred parts, it shall still in the same manner continue to float: but encreasing its thickness only a Hairs breadth, it will alwaies submerge, although we should multiply the Superficies a hundred and a hundred times. Now forasmuch as that this is a Cause, which being added, we adde also the Effect, and being removed, it is removed; and by augmenting or lessening the length or breadth in any manner, the effect of going, or not going to the bottom, is not added or removed: I conclude, that the greatness and smalness of the Superficies hath no influence upon the Natation or Submersion. And that the proportion of the Altitude of the Ramparts of Water, to the Altitude of the Solid, being constituted in the manner aforesaid, the greatness or smalness of the Superficies, makes not any variation, is manifest from that which hath been above demonstrated, and from this, that, *The Prisms and Cylinders which have the same Base, are in proportion to one another as their heights.* Whence Cylinders or Prismes, namely, the Board, be they great or little, so that they be all of equall thickness, have the same proportion to their Conterminall Air, which hath for Base the said Superficies of the Board, and for height the Ramparts of water; so that alwayes of that Air, and of the Board, Solids, are compounded, that in Gravity equall a Mass of water equall to the Mass of the Solids, compounded of Air, and of the Board: whereupon all the said Solids do in the same manner continue afloat. We will conclude in the third place, that,

THEOREME. IX.

All Figures of all Matters, float by hep of the Rampart replenished with Air, and some but only touch the water.

All sorts of Figures of whatsoever Matter, albeit more grave than the Water, do by Benefit of the said Rampart, not only float, but some Figures, though of the gravest Matter, do stay wholly above Water, wetting only the inferiour Surface that toucheth the Water.

And these shall be all Figures, which from the inferiour Base upwards, grow lesser and lesser; the which we shall exemplifie for this time in Piramides or Cones, of which Figures the passions are common. We will demonstrate therefore, that,

It is possible to form a Piramide, of any whatsoever Matter preposed, which being put with its Base upon the Water, rests not only without submerging, but without wetting it more then its Base.

For the explication of which it is requisite, that we first demonstrate the subsequent Lemma, namely, that,

LEMMA II.

Solids whose Masses answer in proportion contrarily to their Specificall Gravities, are equall in Absolute Gravities.

Let A C and B be two Solids, and let the Mass A C be to the Mass B, as the Specificall Gravity of the Solid B, is to the Specificall Gravity of the Solid A C: I say, the Solids A C and B are equall in

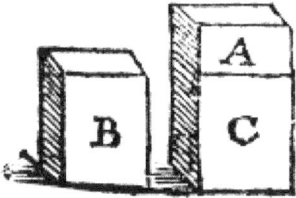

absolute weight, that is, equally grave. For if the Mass A C be equall to the Mass B, then, by the Assumption, the Specificall Gravity of B, shall be equall to the Specificall Gravity of A C, and being equall in Mass, and of the same Specificall Gravity they shall absolutely weigh one as much as another. But if their Masses shall be unequall, let the Mass A C be greater, and in it take the part C, equall to the Mass B. And, because the Masses B and C are equall; the Absolute weight of B, shall have the same proportion to the Absolute weight of C, that the Specificall Gravity of B, hath to the Specificall Gravity of C; or of C A, which is the same *in specie*: But look what proportion the Specificall Gravity of B, hath to the Specificall Gravity of C A, the like proportion, by the Assumption, hath the Mass C A, to the Mass B, that is, to the Mass C: Therefore, the absolute weight of B, to the absolute weight of C, is as the Mass A C to the Mass C: But as the Mass A C, is to the Mass C, so is the absolute weight of A C, to the absolute weight of C: Therefore the absolute weight of B, hath the same proportion to the absolute weight of C, that the absolute weight of A C, hath to the absolute weight of C: Therefore, the two Solids A C and B are equall in absolute Gravity: which was to be demonstrated. Having demonstrated this, I say,

THEOREME X.

That it is possible of any assigned Matter, to form a Piramide or Cone upon any Base, which being put upon the Water shall not submerge, nor wet any more than its Base.

There may be Cones and Piramides of any Matter, which demitted into the water, rest only their Bases.

Let the greatest possible Altitude of the Rampart be the Line D B, and the Diameter of the Base of the Cone to be made of any Matter assigned B C, at right angles to D B: And as the Specificall Gravity

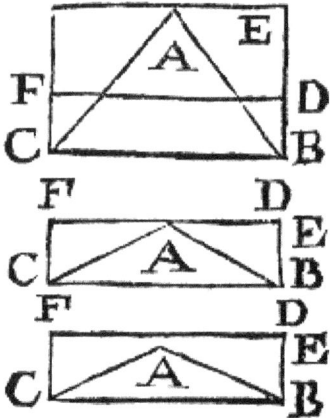

of the Matter of the Piramide or Cone to be made, is to the Specificall Gravity of the water, so let the Altitude of the Rampart D B, be to the third part of the Piramide or Cone A B C, described upon the Base, whose Diameter is B C: I say, that the said Cone A B C, and any other Cone, lower then the same, shall rest upon the Surface of the water B C without sinking. Draw D F parallel to B C, and suppose the Prisme or Cylinder E C, which shall be tripple to the Cone A B C. And, because the Cylinder D C hath the same proportion to the Cylinder C E, that the Altitude D B, hath to the Altitude B E: But the Cylinder C E, is to the Cone A B C, as the Altitude E B is to the third part of the Altitude of the Cone: Therefore, by Equality of proportion, the Cylinder D C is to the Cone A B C, as D B is to the third part of the Altitude B E: But as D B is to the third part of B E, so is the Specificall Gravity of the Cone A B C, to the Specificall Gravity of the water: Therefore, as the Mass of the Solid D C, is to the Mass of the Cone A *B* C, so is the Specificall Gravity of the said Cone, to the Specificall Gravity of the water: Therefore, by the precedent Lemma, the Cone A B C weighs in absolute Gravity, as much as a Mass of Water equall to the Mass D C: But the water which by the imposition of the Cone A B C, is driven out of its place, is as much as would precisely lie in the place D C, and is equall in weight to the Cone that displaceth it: Therefore, there shall be an *Equilibrium*, and the Cone shall rest without farther submerging. And its manifest,

COROLARY I.

That making upon the same Basis, a Cone of a less Altitude, it shall be also less grave, and shall so much the more rest without Submersion

Amongst Cones of the same Base, those of least Altitude shall sink the least.

COROLARY II.

It is manifest, also, that one may make Cones and Piramids of any Matter whatsoever, more grave than the water, which being put into the water, with the Apix or Point downwards, rest without Submersion.

There may be Cones and Piramides of any Matter, which demitted with the Point downwards do float atop.

Because if we reassume what hath been above demonstrated, of Prisms and Cylinders, and that on Bases equall to those of the said Cylinders, we make Cones of the same Matter, and three times as high as the Cylinders, they shall rest afloat, for that in Mass and Gravity they shall be equall to those Cylinders, and by having their Bases equall to those of the Cylinders, they shall leave equall Masses of Air included within the Ramparts. This, which for Example sake hath been demonstrated, in Prisms, Cylinders, Cones and Piramids, might be proved in all other Solid Figures, but it would require a whole Volume (such is the multitude and variety of their Symptoms and Accidents) to comprehend the particuler demonstration of them all, and of their severall Segments: but I will to avoid prolixity in the present Discourse, content my self, that by what I have declared every one of ordinary Capacity may comprehend, that there is not any Matter so grave, no not Gold it self, of which one may not form all sorts of Figures, which by vertue of the superiour Air adherent to them, and not by the Waters Resistance of Penetration, do remain afloat, so that they sink not. Nay, farther, I will shew, for removing that Error, that,

THEOREME XI.

A Piramide or Cone put into the Water, with the Point downward shall swimme, and the same put with the Base downwards shall sinke, and it shall be impossible to make it float.

A Piramide or Cone, demitted with the Point downwards shal swim, with its Base downward shall sink.

Now the quite contrary would happen, if the difficulty of Penetrating the water, were that which had hindred the descent, for that the said Cone is far apter to pierce and penetrate with its sharp Point, than with its broad and spacious Base.

And, to demonstrate this, let the Cone be *A B C*, twice as grave as the water, and let its height be tripple to the height of the Rampart *D A E C*: I say, first, that being put lightly into the water with

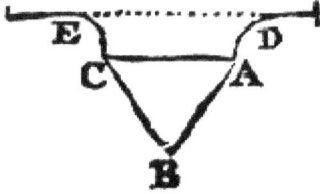

the Point downwards, it shall not descend to the bottom: for the Aeriall Cylinder contained betwixt the Ramparts *D A C E*, is equall in Mass to the Cone *A B C*; so that the whole Mass of the Solid compounded of the Air *D A C E*, and of the Cone *A B C*, shall be double to the Cone *A C B*: And, because the Cone *A B C* is supposed to be of Matter double in Gravity to the water, therefore as much water as the whole Masse *D A B C E*, placed beneath the Levell of the water, weighs as much as the Cone *A B C*: and, therefore, there shall be an *Equilibrium*, and the Cone *A B C* shall descend no lower. Now, I say farther, that the same Cone placed with the Base downwards, shall sink to the bottom, without any possibility of returning again, by any means to swimme.

Let, therefore, the Cone be *A B D*, double in Gravity to the water,

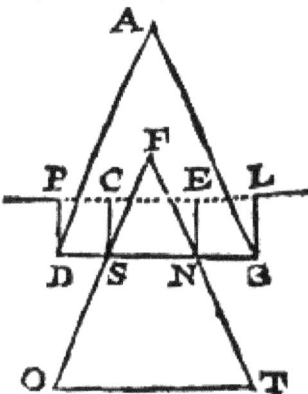

and let its height be tripple the height of the Rampart of water L B: It is already manifest, that it shall not stay wholly out of the water, because the Cylinder being comprehended betwixt the Ramparts *L B D P*, equall to the Cone *A B D*, and the Matter of the Cone, beig double in Gravity to the water, it is evident that the weight of the said Cone shall be double to the weight of the Mass of water equall to the Cylinder *L B D P*: Therefore it shall not rest in this state, but shall descend.

COROLARY I.

I say farther; that much lesse shall the said Cone stay afloat, if one immerge a part thereof.

Much less shall the said Cone swim, if one immerge a part thereof.

Which you may see, comparing with the water as well the part that shall immerge as the other above water. Let us therefore of the Cone A B D, submergeth part N T O S, and advance the Point N S F above water. The Altitude of the Cone F N S, shall either be more than half the whole Altitude of the Cone F T O, or it shall not be more: if it shall be more than half, the Cone F N S shall be more than half of the Cylinder E N S C: for the Altitude of the Cone F N S, shall be more than Sesquialter of the Altitude of the Cylinder E N S C: And, because the Matter of the Cone is supposed to be double in Specificall Gravity to the water, the water which would be contained within the Rampart E N S C, would be less grave absolutely than the Cone F N S; so that the whole Cone F N S cannot be sustained by the Rampart: But the part immerged N T O S, by being double in Specificall Gravity to the water, shall tend to the bottom: Therefore, the whole Cone F T O, as well in respect of the part submerged, as the part above water shall descend to the bottom. But if the Altitude of the Point F N S, shall be half the Altitude of the whole Cone F T O, the same Altitude of the said Cone F N S shall be Sesquialter to the Altitude E N: and, therefore, E N S C shall be double to the Cone F N S; and as much water in Mass as the Cylinder E N S C, would weigh as much as the part of the Cone F N S. But, because the other immerged part N T O S, is double in Gravity to the water, a Mass of water equall to that compounded of the Cylinder E N S C, and of the Solid N T O S, shall weigh less than the Cone F T O, by as much as the weight of a Mass of water equall to the Solid N T O S: Therefore, the Cone sha{l}l also descend. Again, because the Solid N T O S, is septuple to the Cone F N S, to which the Cylinder E

S is double, the proportion of the Solid N T O S, shall be to the *Cy*linder E N S C, as seaven to two: Therefore, the whole Solid compounded of the *C*ylinder E N S C, and of the Solid N T O S, is much less than double the Solid N T O S: Therefore, the single Solid N T O S, is much graver than a Mass of water equall to the Mass, compounded of the *C*ylinder E N S C, and of N T O S.

COROLARY II.

From whence it followeth, that though one should remove and take away the part of the Cone F N S, the sole remainder N T O S would go to the bottom.

Part of the Cones towards the Cuspis removed, it shall still sink.

COROLARY III.

And if we should more depress the Cone F T O, it would be so much the more impossible that it should sustain it self afloat, the part submerged N T O S still encreasing, and the Mass of Air contained in the Rampart diminishing, which ever grows less, the more the Cone submergeth.

The more the Cone is immerged, the more impossible is its floating.

That Cone, therefore, that with its Base upwards, and its *Cuspis* downwards doth swimme, being dimitted with its Base downward must of necessity sinke. They have argued farre from the truth, therefore, who have ascribed the cause of Natation to waters resistance of Division, as to a passive principle, and to the breadth of the Figure, with which the division is to be made, as the Efficient.

I come in the fourth place, to collect and conclude the reason of that which I have proposed to the Adversaries, namely,

THEOREME XII.

That it is possible to fo{r}m Solid Bodies, of what Figure and greatness soever, that of their own Nature goe to the Bottome; But by the help of the Air contained in the Rampart, rest without submerging.

Solids of any Figure & greatnesse, that naturally sink, may by help of the Air in the Rampart swimme.

The truth of this Proposition is sufficiently manifest in all those Solid Figures, that determine in their uppermost part in a plane Superficies: for making such Figures of some Matter specifically as grave as the water, putting them into the water, so that the whole Mass be covered, it is manifest, that they shall rest in all places, provided, that such a Matter equall in weight to the water, may be exactly adjusted: and they shall by consequence, rest or lie even with the Levell of the water, without making any Rampart. If, therefore, in respect of the Matter, such Figures are to rest without submerging, though deprived of the help of the Rampart, it is manifest, that they may admit so much encrease of Gravity, (without encreasing their Masses) as is the weight of as much water as would be contained within the Rampart, that is made about their upper plane Surface: by the help of which being sustained, they shall rest afloat, but being bathed, they shall descend, having been made graver than the water. In Figures, therefore, that determine above in a plane, we may cleerly comprehend, that the Rampart added or removed, may prohibit or permit the descent: but in those Figures that go lessening upwards towards the top, some Persons may, and that not without much seeming Reason, doubt whether the same may be done, and especially by those which terminate in a very acute Point, such as are your Cones and small Piramids. Touching these, therefore, as more dubious than the rest, I will endeavour to demonstrate, that they also lie under the same Accident of going, or not going to the Bottom, be they of any whatever bigness. Let therefore the Cone be A B D, made of a matter

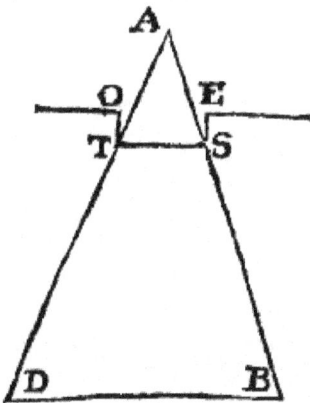

specifically as grave as the water; it is manifest that being put all under water, it shall rest in all places (alwayes provided, that it shall weigh exactly as much as the water, which is almost impossible to effect) and that any small weight being added to it, it shall sink to the bottom: but if it shall descend downwards gently, I say, that it shall make the Rampart E S T O, and that there shall stay out of the water the point A S T, tripple in height to the Rampart E S: which is manifest, for the Matter of the Cone weighing equally with the water, the part submerged *S B D T*, becomes indifferent to move downwards or upwards; and the Cone *A S T*, being equall in Mass to the water that would be contained in the concave of the Rampart *E S T O*, shall be also equall unto it in Gravity: and, therefore, there shall be a perfect *Equilibrium*, and, consequently, a Rest. Now here ariseth a doubt, whether the Cone *A B D* may be made heavier, in such sort, that when it is put wholly under water, it goes to the bottom, but yet not in such sort, as to take from the Rampart the vertue of sustaining it that it sink not, and, the reason of the doubt is this: that although at such time as the Cone *A B D* is specifically as grave as the water, the Rampart *E S T O* sustaines it, not only when the point *A S T* is tripple in height to the Altitude of the Rampart *E S*, but also when a lesser part is above water; [for although in the Descent of the Cone the

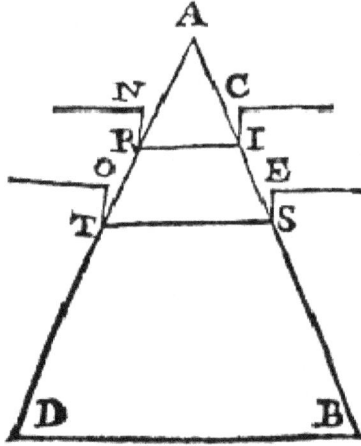

Point *A S T* by little and little diminisheth, and so likewise the Rampart *E S T O*, yet the Point diminisheth in greater proportion than the Rampart, in that it diminisheth according to all the three Dimensions, but the Rampart according to two only, the Altitude still remaining the same; or, if you will, because the Cone *S {A} T* goes diminishing, according to the proportion of the cubes of the Lines that do successively become the Diameters of the Bases of emergent Cones, and the Ramparts diminish according to the proportion of the Squares of the same Lines; whereupon the proportions of the Points are alwayes Sesquialter of the proportions of the Cylinders, contained within the Rampart; so that if, for Example, the height of the emergent Point were double, or equall to the height of the Rampart, in these cases, the Cylinder contained within the Rampart, would be much greater than the said Point, because it would be either sesquialter or tripple, by reason of which it would perhaps serve over and above to sustain the whole Cone, since the part submerged would no longer weigh any thing;] yet, nevertheless, when any Gravity is added to the whole Mass of the Cone, so that also the part submerged is not without some excesse of Gravity above the Gravity of the water, it is not manifest, whether the Cylinder contained within the Rampart, in the descent that the Cone shall make, can be reduced to such a proportion unto the emergent Point, and to such an excesse of Mass above the Mass of it, as to compensate the excesse of the Cones Specificall Gravity above the Gravity of the water: and the Scruple ariseth, because that howbeit in the descent made by the Cone, the emergent Point $A\ S\ T$ diminisheth, whereby there is also a diminution of the excess of the Cones Gravity above the Gravity of the water, yet the case stands so, that the Rampart doth also contract it self, and the Cylinder contained in it doth deminish. Nevertheless it shall be demonstrated, how that the Cone $A\ B\ D$ being of any supposed bignesse, and made at the first of a Matter exactly equall in Gravity to the Water, if there may be affixed to it some Weight, by means of which i{t} may descend to the bottom, when submerged under water, it may also by vertue of the Rampart stay above without sinking.

Let, therefore, the Cone $A\ B\ D$ be of any supposed greatnesse, and alike in specificall Gravity to the water. It is manifest, that being put lightly into the water, it shall rest without descending; and it

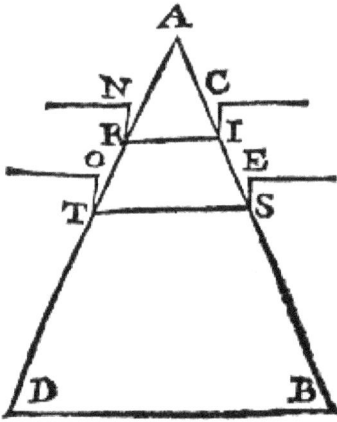

shall advance above water, the Point *A S T*, tripple in height to the height of the Rampart *E S*: Now, suppose the Cone *A B D* more depressed, so that it advance above water, only the Point *A I R*, higher by half than the Point *A S T*, with the Rampart about it *C I R N*. And, because, the Cone *A B D* is to the Cone *A I R*, as the cube of the Line *S T* is to the cube of the Line *I R*, but the Cylinder *E S T O*, is to the Cylinder *C I R N*, as the Square of *S T* to the Square of *I R*, the Cone *A S T* shall be Octuple to the Cone *A I R*, and the Cylinder *E S T O*, quadruple to the Cylinder *C I R N*: But the Cone *A S T*, is equall to the Cylinder *E S T O*: Therefore, the Cylinder *C I R N*, shall be double to the Cone *A I R*: and the water which might be contained in the Rampart *C I R N*, would be double in Mass and in Weight to the Cone *A I R*, and, therefore, would be able to sustain the double of the Weight of the Cone *A I R*: Therefore, if to the whole Cone *A B D*, there be added as much Weight as the Gravity of the Cone *A I R*, that is to say, the eighth part of the weight of the Cone *A S T*, it also shall be sustained by the Rampart *C I R N*, but without that it shall go to the bottome: the Cone *A B D*, being, by the addition of the eighth part of the weight of the Cone *A S T*, made specifically more grave than the water. But if the Altitude of the Cone *A I R*, were two thirds of the Altitude of the Cone *A S T*, the Cone *A S T* would be to the Cone *A I R*, as twenty seven to eight; and the Cylinder *E S T O*, to the Cylinder *C I R N*, as nine to four, that is, as twenty seven to twelve; and, therefore, the Cylinder *C I R N*, to the Cone *A I R*, as twelve to eight; and the excess of the Cylinder *C I R N*, above the Cone *A I R*, to the Cone *A S T*, as four to twenty seven: therefore if to the Cone *A B D* be added so much weight as is the four twenty sevenths of the weight of the Cone *A S T*, which is a little more then its seventh part, it also shall continue to swimme, and the height of the emergent Point shall be double to the height of the Rampart. This that hath been demonstrated in Cones, exactly holds in Piramides, although the one or the other should be very sharp in their Point or Cuspis: From whence we conclude, that the same Accident shall so much the more easily happen in all other Figures, by how much the less sharp the Tops shall be, in which they determine, being assisted by more spacious Ramparts.

THEOREME XIII.

All Figures, therefore, of whatever greatnesse, may go, and not go, to the Bottom, according as their Sumities or Tops shall be bathed or not bathed.

All Figures sink or swim, upon bathing or not bathing of their tops.

And this Accident being common to all sorts of Figures, without exception of so much as one. Figure hath, therefore, no part in the production of this Effect, of sometimes sinking, and sometimes again not sinking, but only the being sometimes conjoyned to, and sometimes seperated from, the supereminent Air: which cause, in fine, who so shall rightly, and, as we say, with both his Eyes, consider this business, will find that it is reduced to, yea, that it really is the same with, the true, Naturall and primary cause of Natation or Submersion; to wit, the excess or deficiency of the Gravity of the water, in relation to the Gravity of that Solid Magnitude, that is demitted into the water. For like as a Plate of Lead, as thick as the back of a Knife, which being put into the water by it self alone goes to the bottom, if upon it you fasten a piece of Cork four fingers thick, doth continue afloat, for that now the Solid that is demitted in the water, is not, as before, more grave than the water, but less, so the Board of Ebony, of its own nature more grave than water; and, therefore, descending to the bottom, when it is demitted by it self alone into the water, if it shall be put upon the water, conjoyned with an Expanded vail of Air, that together with the Ebony doth descend, and that it be such, as that it doth make with it a compound less grave than so much water in Mass, as equalleth the Mass already submerged and depressed beneath the Levell of the waters Surface, it shall not descend any farther, but shall rest, for no other than the universall and most common cause, which is that Solid Magnitudes, less grave *in specie* than the water, go not to the bottom.

So that if one should take a Plate of Lead, as for Example, a finger thick, and an handfull broad every way, and should attempt to make it swimme, with putting it lightly on the water, he would lose his Labour, because that if it should be depressed an Hairs breadth beyond the possible Altitude of the Ramparts of water, it would dive and sink; but if whilst it is going downwards, one should make certain Banks or Ramparts about it, that should hinder the defusion of the water upon the said Plate, the which Banks should rise so high, as that they might be able to contain as much water, as should weigh equally with the said Plate, it would, witho{u}t all Question, descend no lower, but would rest, as being sustained by vertue of the Air contained within the aforesaid Ramparts: and, in short, there would be a Vessell by this means formed with the bottom of Lead. But if the thinness of the Lead shall be such, that a very small height of Rampart would suffice to contain so much Air, as might keep it afloat, it shall also rest without the Artificiall Banks or Ramparts, but yet not without the Air, because the Air by it self makes Banks sufficient for a small height, to resist the Superfusion of the water: so that that which in this case swimmes, is as it were a Vessell filled with Air, by vertue of which it continueth afloat.

I will, in the last place, with an other Experime{n}t, attempt to remove all difficulties, if so be there should yet be any doubt left in any one, touching the opperation of this Continuity[5] of the Air, with the thin Plate which swims, and afterwards put an end to this part of my discourse.

I suppose my self to be questioning with some of my Oponents.

Whether Figure have any influence upon the encrease or diminution of the Resistance in any Weight against its being raised in the Air; and I suppose, that I am to maintain the Affirmative, asserting that a Mass of Lead, reduced to the Figure of a Ball, shall be raised with less force, then if the same had been made into a thinne and broad Plate, because that it in this spacious Figure, hath a great quantity of Air to penetrate, and in that other, more compacted and contracted very little: and to demonstrate the truth of such my Opinion, I will hang in a small thred first the Ball or Bullet, and put that into the water, tying the thred that upholds it to one end of the Ballance that I hold in the Air, and to the other end I by degrees adde so much Weight, till that at last it brings

---

[5] Or rather Contiguity

up the Ball of Lead out of the water: to do which, suppose a Gravity of thirty Ounces sufficeth; I afterwards reduce the said Lead into a flat and thinne Plate, the which I likewise put into the water, suspended by three threds, which hold it parallel to the Surface of the water, and putting in the same manner, Weights to the other end, till such time as the Plate comes to be raised and drawn out of the water: I finde that thirty six ounces will not suffice to seperate it from the water, and raise it thorow the Air: and arguing from this Experiment, I affirm, that I have fully demonstrated the truth of my Proposition. Here my Oponents desires me to look down, shewing me a thing which I had not before observed, to wit, that in the Ascent of the Plate out of the water, it draws after it another Plate (*if I may so call it*) of water, which before it divides and parts from the inferiour Surface of the Plate of Lead, is raised above the Levell of the other water, more than the thickness of the back of a Knife: Then he goeth to repeat the Experiment with the Ball, and makes me see, that it is but a very small quantity of water, which cleaves to its compacted and contracted Figure: and then he subjoynes, that its no wonder, if in seperating the thinne and broad Plate from the water, we meet with much greater Resistance, than in seperating the Ball, since together with the Plate, we are to raise a great quantity of water, which occurreth not in the Ball: He telleth me moreover, how that our Question is, whether the Resistance of Elevation be greater in a dilated Plate of Lead, than in a Ball, and not whether more resisteth a Plate of Lead with a great quantity of water, or a Ball with a very little water: He sheweth me in the close, that the putting the Plate and the Ball first into the water, to make proofe thereby of their Resistance in the Air, is besides our case, which treats of Elivating in the Air, and of things placed in the Air, and not of the Resistance that is made in the Confines of the Air and water, and by things which are part in Air and part in water: and lastly, they make me feel with my hand, that when the thinne Plate is in the Air, and free from the weight of the water, it is raised with the very same Force that raiseth the Ball. Seeing, and understanding these things, I know not what to do, unless to grant my self convinced, and to thank such a Friend, for having made me to see that which I never till then observed: and, being advertised by this same Accident, to tell my Adversaries, that our Question is, whether a Board and a Ball of Ebony, equally go to the bottom in water, and not a Ball of Ebony and a Board of Ebony, joyned with another flat Body of Air: and, farthermore, that we speak of sinking, and not sinking to the bottom, in water, and not of that which happeneth in the Confines of the water and Air to Bodies that be part in the Air, and part in the water; nor much less do we treat of the greater or lesser Force requisite in seperating this or that Body from the Air; not omitting to tell them, in the last place, that the Air doth resist, and gravitate downwards in the water, just so much as the water (if I may so speak) gravitates and resists upwards in the Air, and that the same Force is required to sinke a Bladder under water, that is full of Air, as to raise it in the Air, being full of water, removing the consideration of the weight of that Filme or Skinne, and considering the water and the Air only. And it is likewise true, that the same Force is required to sink a Cup or such like Vessell under water, whilst it is full of Air, as to raise it above the Superficies of the water, keeping it with the mouth downwards; whilst it is full of water, which is constrained in the same manner to follow the Cup which contains it, and to rise above the other water into the Region of the Air, as the Air is forced to follow the same Vessell under the Surface of the water, till that in this c{a}se the water, surmounting the brimme of the Cup, breaks in, driving thence the Air, and in that case, the said brimme coming out of the water, and arriving to the Confines of the Air, the water falls down, and the Air sub-enters to fill the cavity of the Cup: upon which ensues, that he no less transgresses the Articles of the *Convention*, who produceth a Plate conjoyned with much Air, to see if it descend to the bottom in water, then he that makes proof of the Resistance against Elevation in Air with a Plate of Lead, joyned with a like quantity of water.

Aristotles opinion touching the Operation of Figure examined.

I have said all that I could at present think of, to maintain the Assertion I have undertook. It remains, that I examine that which *Aristotle* hath writ of this matter towards the end of his Book De Cælo; wherein I shall note two things: the one that it being true as hath been demonstrated, that Figure hath nothing to do about the moving or not moving it self upwards or downwards, its seemes that *Aristotle* at his first falling upon this Speculation, was of the same opinion, as in my opinion may be collected from the examination of his words. 'Tis true, indeed, that in essaying afterwards to render a reason of such effect, as not having in my conceit hit upon the right, (which in the second place I will examine) it seems that he is brought to admit the largenesse of Figure, to be interested in this operation. As to the first particuler, hear the precise words of *Aristotle*.

Figures are not the Causes of moving simply upwards or downwards, but of moving more slowly or swiftly, and by what means this comes to pass, it is not difficult to see.

Here first I note, that the terms being four, which fall under the present consideration, namely, Motion, Rest, Slowly and Swiftly: And *Aristotle* naming figures as Causes of Tardity and Velocity, excluding them from being the Cause of absolute and simple Motion, it seems necessary, that he exclude them on the other side, from being the Cause of Rest, so that his meaning is this. Figures are not the Causes of moving or not moving absolutely, but of moving quickly or slowly: and, here, if any should say the mind of *Aristotle* is to exclude Figures from being Causes of Motion, but yet not from being Causes of Rest, so that the sence would be to remove from Figures, there being the Causes of moving simply, but yet not there being Causes of Rest, I would demand, whether we ought with *Aristotle* to understand, that all Figures universally, are, in some manner, the causes of Rest in those Bodies, which otherwise would move, or else some particular Figures only, as for Example, broad and thinne Figures: If all indifferently, then every Body shall rest: because every Body hath some Figure, which is false; but if some particular Figures only may be in some manner a Cause of Rest, as, for Example, the broad, then the others would be in some manner the Causes of Motion: for if from seeing some Bodies of a contracted Figure move, which after dilated into Plates rest, may be inferred, that the Amplitude of Figure hath a part in the Cause of that Rest; so from seeing such like Figures rest, which afterwards contracted move, it may with the same reason be affirmed, that the united and contracted Figure, hath a part in causing Motion, as the remover of that which impeded it: The which again is directly opposite to what *Aristotle* saith, namely, that Figures are not the Causes of Motion. Besides, if *Aristotle* had admitted and not excluded Figures from being Causes of not moving in some Bodies, which moulded into another Figure would move, he would have impertinently propounded in a dubitative manner, in the words immediately following, whence it is, that the large and thinne Plates of Lead or Iron, rest upon the water, since the Cause was apparent, namely, the Amplitude of Figure. Let us conclude, therefore, that the meaning of *Aristotle* in this place is to affirm, that Figures are not the Causes of absolutely moving or not moving, but only of moving swiftly or slowly: which we ought the rather to believe, in regard it is indeed a most true conceipt and opinion. Now the mind of *Aristotle* being such, and appearing by consequence, rather contrary at the first sight, then favourable to the assertion of the Oponents, it is necessary, that their Interpretation be not exactly the same with that, but such, as being in part understood by some of them, and in part by others, was set down: and it may easily be indeed so, being an Interpretation consonent to the sence of the more famous Interpretors, which is, that the Adverbe *Simply* or *Absolutely*, put in the Text, ought not to be joyned to the Verbe to *Move*, but with the Noun *Causes*: so that the purport of *Aristotles* words, is to affirm, That Figures are not the Causes absolutely of moving or not moving, but yet are Causes *Secundum quid, viz.* in some sort; by which means, they are called Auxiliary and Concomitant Causes: and this Proposition is received and asserted as true by *Signor Buonamico Lib. 5. Cap. 28.* where he thus writes. *There are other Causes concomitant, by which some things float, and others sink, among which the Figures of Bodies hath the first place*, &c.

Concerning this Proposition, I meet with many doubts and difficulties, for which me thinks the words of *Aristotle* are not capable of such a construction and sence, and the difficulties are these.

First in the order and disposure of the words of *Aristotle*, the particle *Simpliciter*, or if you will *absoluté*, is conjoyned with the Verb *to move*, and seperated from the Noun *Causes*, the which is a great presumption in my favour, seeing that the writing and the Text saith, Figures are not the Cause of moving simply upwards or downwards, but of quicker or slower Motion: and, saith not, Figures are not simply the Causes of moving upwards or downwards, and when the words of a Text receive, transposed, a sence different from that which they sound, taken in the order wherein the Author disposeth them, it is not convenient to inverte them. And who will affirm that *Aristotle* desiring to write a Proposition, would dispose the words in such sort, that they should import a different, nay, a contrary sence? contrary, I say, because understood as they are written; they say, that Figures are not the Causes of Motion, but inverted, they say, that Figures are the Causes of Motion, &c.

Moreover, if the intent of *Aristotle* had been to say, that Figures are not simply the Causes of moving upwards or downwards, but only Causes *Secundum quid*, he would not have adjoyned those words, *but they are Causes of the more swift or slow Motion*; yea, the subjoining this would have been not only superfluous but false, for that the whole tenour of the Proposition would import thus much. Figures are not the absolute Causes of moving upwards or downwards, but are the absolute Cause of the swift or slow Motion; which is not true: because the primary Causes of greater or lesser Velocity, are by *Aristotle* in the 4th of his *Physicks, Text. 71.* attributed to the greater or lesser Gravity of Moveables, compared among themselves, and to the greater or lesser Resistance of the *Medium's*, depending on their greater or less Crassitude: and these are inserted by *Aristotle* as the primary Causes; and these two only are in that place nominated: and Figure comes afterwards to be considered, *Text. 74.* rather as an Instrumentall Cause of the force of the Gravity, the which divides either with the Figure, or with the *Impetus*; and, indeed, Figure by it self without the force of Gravity or Levity, would opperate nothing.

I adde, that if *Aristotle* had an opinion that Figure had been in some sort the Cause of moving or not moving, the inquisition which he makes immediately in a doubtfull manner, whence it comes, that a Plate of Lead flotes, would have been impertinent; for if but just before he had said, that Figure was in a certain sort the Cause of moving or not moving, he needed not to call in Question, by what Cause the Plate of Lead swims, and then ascribing the Cause to its Figure; and framing a discourse in this manner. Figure is a Cause *Secundum quid* of not sinking: but, now, if it be doubted, for what Cause a thin Plate of Lead goes not to the bottom; it shall be answered, that that proceeds from its Figure: a discourse which would be indecent in a Child, much more in *Aristotle*; For where is the occasion of doubting? And who sees not, that if *Aristotle* had held, that Figure was in some sort a Cause of Natation, he would without the least Hesitation have writ; That Figure is in a certain sort the Cause of Natation, and therefore the Plate of Lead in respect of its large and expatiated Figure swims; but if we take the proposition of *Aristotle* as I say, and as it is written, and as indeed it is true, the ensuing words come in very oppositely, as well in the introduction of swift and slow, as in the question, which very pertinently offers it self, and would say thus much.

Figures are not the Cause of moving or not moving simply upwards or downwards, but of moving more quickly or slowly: But if it be so, the Cause is doubtfull, whence it proceeds, that a Plate of Lead or of Iron broad and thin doth swim, &c. And the occasion of the doubt is obvious, because it seems at the first glance, that the Figure is the Cause of this Natation, since the same Lead, or a less quantity, but in another

Figure, goes to the bottom, and we have already affirmed, that the Figure hath no share in this effect.

Lastly, if the intent of *Aristotle* in this place had been to say, that Figures, although not absolutely, are at least in some measure the Cause of moving or not moving: I would have it considered, that he names no less the Motion upwards, than the other downwards: and because in exemplifying it afterwards, he produceth no other Experiments than of a Plate of Lead, and Board of Ebony, Matters that of their own Nature go to the bottom, but by vertue (as our Adversaries say) of their Figure, rest afloat; it is fit that they should produce some other Experiment of those Matters, which by their Nature swims, but retained by their Figure rest at the bottom. But since this is impossible to be done, we conclude, that *Aristotle* in this place, hath not attributed any action to the Figure of simply moving or not moving.

But though he hath exquisitely Philosophiz'd, in investigating the solution of the doubts he proposeth, yet will I not undertake to maintain, rather various difficulties, that present themselves unto me, give me occasion of suspecting that he hath not entirely displaid unto us, the true Cause of the present Conclusion: which difficulties I will propound one by one, ready to change opinion, whenever I am shewed, that the Truth is different from what I say; to the confession whereof I am much more inclinable than to contradiction.

Aristotle erred in affirming a Needle dimitted long wayes to sink.

Aristotle having propounded the Question, whence it proceeds, that broad Plates of Iron or Lead, float or swim; he addeth (as it were strengthening the occasion of doubting) forasmuch as other things, less, and less grave, be they round or long, as for instance a Needle go to the bottom. Now I here doubt, or rather am certain that a Needle put lightly upon the water, rests afloat, no less than the thin Plates of Iron or Lead. I cannot believe, albeit it hath been told me, that some to defend *Aristotle* should say, that he intends a Needle demitted not longwayes but endwayes, and with the Point downwards; nevertheless, not to leave them so much as this, though very weak refuge, and which in my judgement *Aristotle* himself would refuse, I say it ought to be understood, that the Needle must be demitted, according to the Dimension named by *Aristotle*, which is the length: because, if any other

Dimension than that which is named, might or ought to be taken, I would say, that even the Plates of Iron and Lead, sink to the bottom, if they be put into the water edgewayes and not flatwayes. But because *Aristotle* saith, broad Figures go not to the bottom, it is to be understood, being demitted broadwayes: and, therefore, when he saith, long Figures as a Needle, albeit light, rest not afloat, it ought to be understood of them when demitted longwayes.

Moreover, to say that Aristotle *is to be understood of the Needle demitted with the Point downwards, is to father upon him a great impertinency; for in this place he saith, that little Particles of Lead or Iron, if they be round or long as a Needle, do sink to the bottome; so that by his Opinion, a Particle or small Grain of Iron cannot swim: and if he thus believed, what a great folly would it be to subjoyn, that neither would a Needle demitted endwayes swim? And what other is such a Needle, but many such like Graines accumulated one upon another? It was too unworthy of such a man to say, that one single Grain of Iron could not swim, and that neither can it swim, though you put a hundred more upon it.*

Lastly, either *Aristotle* believed, that a Needle demitted longwayes upon the water, would swim, or he believed that it would not swim: If he believed it would not swim, he might well speak as indeed he did; but if he believed and knew that it would float, why, together with the dubious Problem of the Natation of broad Figures, though of ponderous Matter, hath he not also introduced the Question; whence it proceeds, that even long and slender Figures, howbeit of Iron or Lead do swim? And the rather, for that the occasion of doubting seems greater in long and narrow Figures, than in broad and thin, as from *Aristotles* not having doubted of it, is manifested.

No lesser an inconvenience would they fasten upon *Aristotle*, who in his defence should say, that he means a Needle pretty thick, and not a small one; for take it for granted to be intended of a small one; and it shall suffice to reply, that he believed that it would swim; and I will again charge him with having avoided a more wonderfull and intricate Probleme, and introduced the more facile and less wonderfull.

We say freely therefore, that *Aristotle* did hold, that only the broad Figure did swim, but the long and slender, such as a Needle, not. The which nevertheless is false, as it is also false in round Bodies: because, as from what hath been predemonstrated, may be gathered, little Balls of Lead and Iron, do in like manner swim.

Aristotle affirmeth some Bodies volatile for their Minuity, Text. 42.

He proposeth likewise another Conclusion, which likewise seems different from the truth, and it is, That some things, by reason of their littleness fly in the Air, as the small dust of the Earth, and the thin leaves of beaten Gold: but in my Opinion, Experience shews us, that that happens not only in the Air, but also in the water, in which do descend, even those Particles or Atomes of Earth, that disturbe it, whose minuity is such, that they are not deservable, save only when they are many hundreds together. Therefore, the dust of the Earth, and beaten Gold, do not any way sustain themselves in the Air, but descend downwards, and only fly to and again in the same, when strong Windes raise them, or other agitations of the Air commove them: and this also happens in the commotion of the water, which raiseth its Sand from the bottom, and makes it muddy. But *Aristotle* cannot mean this impediment of the commotion, of which he makes no mention, nor names other than the lightness of such Minutiæ or Atomes, and the Resistance of the Crassitudes of the Water and Air, by which we see, that he speakes of a calme, and not disturbed and agitated Air: but in that case, neither Gold nor Earth, be they never so small, are sustained, but speedily descend.

Democritus placed the Cause of Natation in certain fiery Atomes.

He passeth next to confute *Democritus*, which, by his Testimony would have it, that some Fiery Atomes, which continually ascend through the water, do spring upwards, and sustain those grave Bodies, which are very broad, and that the narrow descend to the bottom, for that but a small quantity of those Atomes, encounter and resist them.

Democritus confuted by *Aristotle*, text 43[6].

I say, *Aristotle* confutes this position, saying, that that should much more occurre in the Air, as the same *Democritus* instances against

---

[6] *Aristot. De Cælo* lib. 4. cap. 6. text. 43.

himself, but after he had moved the objection, he slightly resolves it, with saying, that those Corpuscles which ascend in the Air, make not their *Impetus* conjunctly. Here I will not say, that the reason alledged by *Democritus* is true, but I will only say, it *Aristotles* confutation of *Democritus* refuted by the Author. seems in my judgement, that it is not wholly confuted by *Aristotle*, whilst he saith, that were it true, that the calid ascending Atomes, should sustain Bodies grave, but very broad, it would much more be done in the Air, than in Water, for that haply in the Opinion of *Aristotle*, the said calid Atomes ascend with much greater Force and Velocity through the Air, than through the water. And if this be so, as I verily believe it is, the Objection of *Aristotle* in my judgement seems to give occasion of suspecting, that he may possibly be deceived in more than one particular: First, because those calid Atomes, (whether they be Fiery Corpuscles, or whether they be Exhalations, or in short, whatever other matter they be, that ascends upwards through the Air) cannot be believed to mount faster through Air, than through water: but rather on the contrary, they peradventure move more impetuously through the water, than through the Air, as hath been in part demonstrated above. And here I cannot finde the reason, why *Aristotle* seeing, that the descending Motion of the same Moveable, is more swift in Air, than in water, hath not advertised us, that from the contrary Motion, the contrary should necessarily follow; to wit, that it is more swift in the water, than in the Air: for since that the Moveable which descendeth, moves swifter through the Air, than through the water, if we should suppose its Gravity gradually to diminish, it would first become such, that descending swiftly through the Air, it would descend but slowly through the water: and then again, it might be such, that descending in the Air, it should ascend in the water: and being made yet less grave, it shall ascend swiftly through the water, and yet descend likewise through the Air: and in short, before it can begin to ascend, though but slowly through the Air, it shall ascend swiftly through the water: how then is it true, that ascending Moveables move swifter through the Air, than through the water?

That which hath made *Aristotle* believe, the Motion of Ascent to be swifter in Air, than in water, was first, the having referred the Causes of slow and quick, as well in the Motion of Ascent, as of Descent, only to the diversity of the Figures of the Moveable, and to the more or less Resistance of the greater or lesser Crassitude, or Rarity of the *Medium*; not regarding the comparison of the Excesses of the Gravities of the Moveables, and of the *Mediums*: the which notwithstanding, is the most principal point in this affair: for if the augmentation and diminution of the Tardity or Velocity, should have only respect to the Density or Rarity of the *Medium*, every Body that descends in Air, would descend in water: because whatever difference is found between the Crassitude of the water, and that of the Air, may well be found between the Velocity of the same Moveable in the Air, and some other Velocity: and this should be its proper Velocity in the water, which is absolutely false. The other occasion is, that he did believe, that like as there is a positive and intrinsecall Quality, whereby Elementary Bodies have a propension of moving towards the Centre of the Earth, so there is another likewise intrinsecall, whereby some of those Bodies have an *Impetus*[7] of flying the Centre, and moving upwards: by Vertue of which intrinsecall Principle, called by him Levity, the Moveables which have that same Motion more easily penetrate the more subtle *Medium*, than the more dense: but such a Proposition appears likewise uncertain, as I have above hinted in part, and as with Reasons and Experiments, I could demonstrate, did not the present Argument importune me, or could I dispatch it in few words.

The Objection therefore of *Aristotle* against *Democritus*, whilst he saith, that if the Fiery ascending Atomes should sustain Bodies grave, but of a distended Figure, it would be more observable in the Air than in the water, because such Corpuscles move swifter in that, than in this, is not good; yea the contrary would evene, for that they ascend more slowly through the Air: and, besides their moving slowly, they ascend, not united together, as in the water, but discontinue, and, as we say, scatter: And, therefore, as *Democritus* well replyes, resolving the instance they make not their push or *Impetus* conjunctly.

Aristotle, in the second place, deceives himself, whilst he will have the said grave Bodies to be more easily sustained by the said Fiery ascending Atomes in the Air than in the Water: not observing, that the said Bodies are much more grave in that, than in this, and that such a Body weighs ten pounds in the Air, which will not in the water weigh 1/2 an ounce; how can it then be more easily sustained in the Air, than in the Water?

Democritus confuted by the Authour.

---

[7] Lib. 4. Cap. 5.

Let us conclude, therefore, that *Democritus* hath in this particular better Philosophated than *Aristotle*. But yet will not I affirm, that *Democritus* hath reason'd rightly, but I rather say, that there is a manifest Experiment that overthrows his Reason, and this it is, That if it were true, that calid ascending Atomes should uphold a Body, that if they did not hinder, would go to the bottom, it would follow, that we may find a Matter very little superiour in Gravity to the water, the which being reduced into a Ball, or other contracted Figure, should go to the bottom, as encountring but few Fiery Atomes; and which being distended afterwards into a dilated and thin Plate, should come to be thrust upwards by the impulsion of a great Multitude of those Corpuscles, and at last carried to the very Surface of the water: which wee see not to happen; Experience shewing us, that a Body *v. gra.* of a Sphericall Figure, which very hardly, and with very great leasure goeth to the bottom, will rest there, and will also descend thither, being reduced into whatsoever other distended Figure. We must needs say then, either that in the water, there are no such ascending Fiery Atoms, or if that such there be, that they are not able to raise and lift up any Plate of a Matter, that without them would go to the bottom: Of which two Positions, I esteem the second to be true, understanding it of water, constituted in its naturall Coldness. But if we take a Vessel of Glass, or Brass, or any other hard matter, full of cold water, within which is put a Solid of a flat or concave Figure, but that in Gravity exceeds the water so little, that it goes slowly to the bottom; I say, that putting some burning Coals under the said Vessel, as soon as the new Fiery Atomes shall have penetrated the substance of the Vessel, they shall without doubt, ascend through that of the water, and thrusting against the foresaid Solid, they shall drive it to the Superficies, and there detain it, as long as the incursions of the said Corpuscles shall last, which ceasing after the removall of the Fire, the Solid being abandoned by its supporters, shall return to the bottom.

But *Democritus* notes, that this Cause only takes place when we treat of raising and sustaining of Plates of Matters, but very little heavier than the water, or extreamly thin: but in Matters very grave, and of some thickness, as Plates of Lead or other Mettal, that same Effect wholly ceaseth: In Testimony of which, let's observe that such Plates, being raised by the Fiery Atomes, ascend through all the depth of the water, and stop at the Confines of the Air, still staying under water: but the Plates of the Opponents stay not, but only when they have their upper Superficies dry, nor is there any means to be used, that when they are within the water, they may not sink to the bottom. The cause, therefore, of the Supernatation of the things of which *Democritus* speaks is one, and that of the Supernatation of the things of which we speak is another. But, returning to *Aristotle*, methinks that he hath more weakly confuted *Democritus*, than *Democritus* himself hath done: For *Aristotle* having propounded the Objection which he maketh against him, and opposed him with saying, that if the calid ascendent Corpuscles were those that raised the thin Plate, much more then would such a Solid be raised and born upwards through the Air, it sheweth that the desire in *Aristotle* to detect *Democritus*, was predominate over the exquisiteness of Solid Philosophizing: which desire of his he hath discovered in other occasions, and that we may not digress too far from this place, in the Text precedent to this Chapter[8] which we have in hand; where he attempts to confute the same *Democritus* for that he, not contenting himself with names only, had essayed more particularly to declare what things Gravity and Levity were; that is, the Causes of descending and ascending, (and had introduced Repletion and Vacuity) ascribing this to Fire, by which it moves upwards, and that to the Earth, by which it descends; afterwards attributing to the Air more of Fire, and to the water more of Earth. But *Aristotle* desiring a positive Cause, even of ascending Motion, and not as *Plato*, or these others, a simple negation, or privation, such as Vacuity would be in reference to Repletion, argueth against *Democritus* and saith: If it be true, as you suppose, then there shall be a great Mass of water, which shall have more of Fire, than a small Mass of Air, and a great Mass of Air, which shall have more of Earth than a little Mass of water, whereby it would ensue, that a great Mass of Air, should come more swiftly downwards, than a little quantity of water: But that is never in any case soever: Therefore *Democritus* discourseth erroneously.

But in my opinion, the Doctrine of *Democritus* is not by this allegation overthrown, but if I erre not, the manner of *Aristotle* deduction either concludes not, or if it do conclude any thing, it may with equall force be restored against himself. *Democritus* will grant to *Aristotle*, that there may be a great Mass of Air taken, which contains more Earth, than a small quantity of water, but yet will deny, that such a Mass of Air, shall go faster downwards than a little water, and that for many reasons. First, because if the greater quantity of Earth, contained in the

---

[8] Cap. 5. Text 41.

great Mass of Air, ought to cause a greater Velocity than a less quantity of Earth, contained in a little quantity of water, it would be necessary, first, that it were true, that a greater Mass of pure Earth, should move more swiftly than a less: But this is false, though *Aristotle* in many places affirms it to be true: because not the greater absolute, but the greater specificall Gravity, is the cause of greater Velocity: nor doth a Ball of Wood, weighing ten pounds, descend more swiftly than one weighing ten Ounces, and that is of the same Matter: but indeed a Bullet of Lead of four Ounces, descendeth more swiftly than a Ball of Wood of twenty Pounds: because the Lead is more *grave in specie* than the Wood. Therefore, its not necessary, that a great Mass of Air, by reason of the much Earth contained in it, do descend more swiftly than a little Mass of water, but on the contrary, any whatsoever Mass of water, shall move more swiftly than any other of Air, by reason the participation of the terrene parts *in specie* is greater in the water, than in the Air. Let us note, in the second place, how that in multiplying the Mass of the Air, we not only multiply that which is therein of terrene, but its Fire also: whence the Cause of ascending, no less encreaseth, by vertue of the Fire, than that of descending on the account of its multiplied Earth. It was requisite in increasing the greatness of the Air, to multiply that which it hath of terrene only, leaving its Fire in its first state, for then the terrene parts of the augmented Air, overcoming the terrene parts of the small quantity of water, it might with more probability have been pretended, that the great quantity of Air, ought to descend with a greater *Impetus*, than the little quantity of water.

Therefore, the Fallacy lyes more in the Discourse of *Aristotle*, than in that of *Democritus*, who with severall other Reasons might oppose *Aristotle*, and alledge; If it be true, that the extreame Elements be one simply grave, and the other simply light, and that the mean Elements participate of the one, and of the other Nature; but the Air more of Levity, and the water more of Gravity, then there shall be a great Mass of Air, whose Gravity shall exceed the Gravity of a little quantity of water, and therefore such a Mass of Air shall descend more swiftly than that little water: But that is never seen to occurr: Therefore its not true, that the mean Elements do participate of the one, and the other quality. This argument is fallacious, no less than the other against *Democritus*.

Lastly, *Aristotle* having said, that if the Position of *Democritus* were true, it would follow, that a great Mass of Air should move more swiftly than a small Mass of water, and afterwards subjoyned, that that is never seen in any Case: methinks others may become desirous to know of him in what place this should evene, which he deduceth against *Democritus*, and what Experiment teacheth us, that it never falls out so. To suppose to see it in the Element of water, or in that of the Air is vain, because neither doth water through water, nor Air through Air move, nor would they ever by any whatever participation others assign them, of Earth or of Fire: the Earth, in that it is not a Body fluid, and yielding to the mobility of other Bodies, is a most improper place and *Medium* for such an Experiment: *Vacuum*, according to the same *Aristotle* himself, there is none, and were there, nothing would move in it: there remains the Region of Fire, but being so far distant from us, what Experiment can assure us, or hath assertained *Aristotle* in such sort, that he should as of a thing most obvious to sence, affirm what he produceth in confutation of *Democritus*, to wit, that a great Mass of Air, is moved no swifter than a little one of water? But I will dwell no longer upon this matter, whereon I have spoke sufficiently: but leaving *Democritus*, I return to the Text of *Aristotle*, wherein he goes about to render the true reason, how it comes to pass, that the thin Plates of Iron or Lead do swim on the water; and, moreover, that Gold it self being beaten into thin Leaves, not only swims in water, but flyeth too and again in the Air. He supposeth that of Continualls, some are easily divisible, others not: and that of the easily divisible, some are more so, and some less: and these he affirms we should esteem the Causes[9]. He addes that that is easily divisible, which is well terminated, and the more the more divisible, and that the Air is more so, than the water, and the water than the Earth. And, lastly, supposeth that in each kind, the lesse quantity is easlyer divided and broken than the greater.

Here I note, that the Conclusions of *Aristotle* in generall are all true, but methinks, that he applyeth them to particulars, in which they have no place, as indeed they have in others, as for Example, Wax is more easily divisible than Lead, and Lead than Silver, inasmuch as Wax receives all the terms more easlier than Lead, and Lead than Silver. Its true, moreover, that a little quantity of Silver is easlier divided than a great Mass: and all these Propositions are true, because true it is, that

---

[9] *De Cælo* l. 4. c. 6. t. 44.

in Silver, Lead and Wax, there is simply a Resistance against Division, and where there is the absolute, there is also the respective. But if as well in water as in Air, there be no Renitence against simple Division, how can we say, that the water is easlier divided than the Air? We know not how to extricate our selves from the Equivocation: whereupon I return to answer, that Resistance of absolute Division is one thing, and Resistance of Division made with such and such Velocity is another. But to produce Rest, and to abate the Motion, the Resistance of absolute Division is necessary; and the Resistance of speedy Division, causeth not Rest, but slowness of Motion. But that as well in the Air, as in water, there is no Resistance of simple Division, is manifest, for that there is not found any Solid Body which divides not the Air, and also the water: and that beaten Gold, or small dust, are not able to superate the Resistance of the Air, is contrary to that which Experience shews us, for we see Gold and Dust to go waving to and again in the Air, and at last to descend downwards, and to do the same in the water, if it be put therein, and separated from the Air. And, because, as I say, neither the water, nor the Air do resist simple Division, it cannot be said, that the water resists more than the Air. Nor let any object unto me, the Example of most light Bodies, as a Feather, or a little of the pith of Elder, or water-reed that divides the Air and not the water, and from this infer, that the Air is easlier divisible than the water; for I say unto them, that if they do well observe, they shall see the same Body likewise divide the Continuity of the water, and submerge in part, and in such a part, as that so much water in Mass would weigh as much as the whole Solid[10]. And if they shal yet persist in their doubt, that such a Solid sinks not through inability to divide the water, I will return them this reply, that if they put it under water, and then let it go, they shall see it divide the water, and presently ascend with no less celerity, than that with which it divided the Air in descending: so that to say that this Solid ascends in the Air, but that coming to the water, it ceaseth its Motion, and therefore the water is more difficult to be divided, concludes nothing: for I, on the contrary, will propose them a piece of Wood, or of Wax, which riseth from the bottom of the water, and easily divides its Resistance, which afterwards being arrived at the Air, stayeth there, and hardly toucheth it; whence I may aswell say, that the water is more easier divided than the Air.

---

[10] *Archimed. De Insident. humi* lib. 2. prop. 1.

I will not on this occasion forbear to give warning of another fallacy of these persons, who attribute the reason of sinking or swimming to the greater or lesse Resistance of the Crassitude of the water against Division, making use of the example of an Egg, which in sweet water goeth to the bottom, but in salt water swims; and alledging for the cause thereof, the faint Resistance of fresh water against Division, and the strong Resistance of salt water. But if I mistake not, from the same Experiment, we may aswell deduce the quite contrary; namely, that the fresh water is more dense, and the salt more tenuous and subtle, since an Egg from the bottom of salt water speedily ascends to the top, and divides its Resistance, which it cannot do in the fresh, in whose bottom it stays, being unable to rise upwards. Into such like perplexities, do false Principles Lead men: but he that rightly Philosophating, shall acknowledge the excesses of the Gravities of the Moveables and of the Mediums, to be the Causes of those effects, will say, that the Egg sinks to the bottom in fresh water, for that it is more grave than it, and swimeth in the salt, for that its less grave than it: and shall without any absurdity, very solidly establish his Conclusions.

Therefore the reason totally ceaseth, that *Aristotle* subjoyns in the Text[11] saying; The things, therefore, which have great breadth remain above, because they comprehend much, and that which is greater, is not easily divided. Such discoursing ceaseth, I say, because its not true, that there is in water or in Air any Resistance of Division; besides that the Plate of Lead when it stays, hath already divided and penetrated the Crassitude of the water, and profounded it self ten or twelve times more than its own thickness: besides that such Resistance of Division, were it supposed to be in the water, could not rationally be affirmed to be more in its superiour parts than in the middle, and lower: but if there were any difference, the inferiour should be the more dense, so that the Plate would be no less unable to penetrate the lower, than the superiour parts of the water; nevertheless we see that no sooner do we wet the superiour Superficies of the Board or thin Piece of Wood, but it precipitatly, and without any retension, descends to the bottom.

I believe not after all this, that any (thinking perhaps thereby to defend *Aristotle*) will say, that it being true, that the much water resists more than the little, the said Board being put lower descendeth, because

---

[11] Text 45.

there remaineth a less Mass of water to be divided by it: because if after the having seen the same Board swim in four Inches of water, and also after that in the same to sink, he shall try the same Experiment upon a profundity of ten or twenty fathom water, he shall see the very self same effect. And here I will take occasion to remember, for the removall of an Error that is too common; That that Ship or other whatsoever Body, that on the depth of an hundred or a thousand fathom, swims with submerging only six fathom of its own height, [*or in the Sea dialect, that draws six fathom water*] shall swim in the same manner in water, that hath but six fathom and half an Inch of depth. Nor do I on the other side, think that it can be said, that the superiour parts of the water are the more dense, although a most grave Authour hath esteemed the superiour water in the Sea to be so, grounding his opinion upon its being more salt, than that at the bottom: but I doubt the Experiment, whether hitherto in taking the water from the bottom, the Observatour did not light upon some spring of fresh water there spouting up: but we plainly see on the contrary, the fresh Waters of Rivers to dilate themselves for some miles beyond their place of meeting with the salt water of the Sea, without descending in it, or mixing with it, unless by the intervention of some commotion or turbulency of the Windes.

But returning to *Aristotle*, I say, that the breadth of Figure hath nothing to do in this business more or less, because the said Plate of Lead, or other Matter, cut into long Slices, swim neither more nor less; and the same shall the Slices do, being cut anew into little pieces, because its not the breadth but the thickness that operates in this business. I say farther, that in case it were really true, that the Renitence to Division were the proper Cause of swimming, the Figures more narrow and short, would much better swim than the more spacious and broad, so that augmenting the breadth of the Figure, the facility of supernatation will be deminished, and decreasing, that this will encrease.

And for declaration of what I say, consider that when a thin Plate of Lead descends, dividing the water, the Division and discontinuation is made between the parts of the water, invironing the perimeter or Circumference of the said Plate, and according to the bigness greater or lesser of that circuit, it hath to divide a greater or lesser quantity of water, so that if the circuit, suppose of a Board, be ten Feet in sinking it flatways, it is to make the seperation and division, and to so speak, an incission upon ten Feet of water; and likewise a lesser Board that is four Feet in Perimeter, must make an incession of four Feet. This granted, he that hath any knowledge in Geometry, will comprehend, not only that a Board sawed in many long thin pieces, will much better float than when it was entire, but that all Figures, the more short and narrow they be, shall so much the better swim. Let the Board A B C D be, for Example, eight Palmes long, and five broad, its circuit shall be twenty six Palmes; and so many must the incession be, which it shall make in the water to descend therein: but if we do saw ir, as suppose into eight little pieces, according to the Lines E F, G H, {&}c. making seven Segments, we must adde to the twenty six Palmes of the circuit of the whole Board, seventy others; whereupon the eight little pieces so cut and seperated, have to cut ninty six Palmes of water. And, if

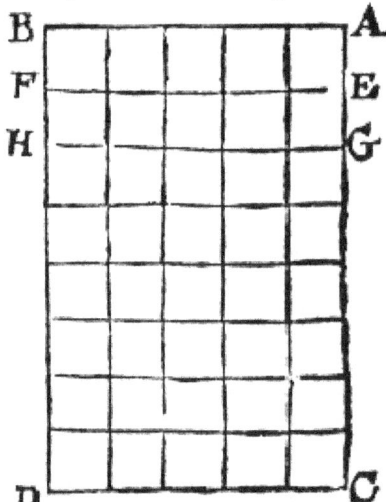

moreover, we **D**   **C** cut each of the said pieces into five parts, reducing them into Squares, to the circuit of ninty six Palmes, with four cuts of eight Palmes apiece; we shall adde also sixty four Palmes, whereupon the said Squares to descend in the water, must divide one hundred and sixty Palmes of water, but the Resistance is much greater than that of twenty six; therefore to the lesser Superficies, we shall reduce them, so much the more easily will they float: and the same will happen in all other Figures, whose Superficies are simular amongst themselves, but different in bigness: because the said Superficies, being either deminished or encreased, always diminish or encrease their Perimeters in subduple proportion; to wit, the Resistance that they find in penetrating the water; therefore the little pieces gradually swim, with more and more facility as their breadth is lessened.

This is manifest; for keeping still the same height of the Solid, with the same proportion as the Base encreaseth or deminisheth, doth the said Solid also encrease or diminish; whereupon the Solid more diminishing than the Circuit, the Cause of Submersion more diminisheth than the Cause of Natation: And on the contrary, the Solid more encreasing than the Circuit, the Cause of Submersion encreaseth more, that of Natation less.

And this may all be deduced out of the Doctrine of *Aristotle* against his own Doctrine.

Lastly, to that which we read in the latter part of the Text[12], that is to say, that we must compare the Gravity of the Moveable with the Resistance of the Medium against Division, because if the force of the Gravity exceed the Resistance of the *Medium*, the Moveable will descend, if not it will float. I need not make any other answer, but that which hath been already delivered; namely, that its not the Resistance of absolute Division, (which neither is in Water nor Air) but the Gravity of the *Medium* that must be compared with the Gravity of the Moveables; and if that of the *Medium* be greater, the Moveable shall not descend, nor so much as make a totall Submersion, but a partiall only; because in the place which it would occupy in the water, there must not remain a Body that weighs less than a like quantity of water: but if the Moveable be more grave, it shall descend to the bottom, and possess a place where it is more conformable for it to remain, than another Body that is less grave. And this is the only true proper and absolute Cause of Natation and Submersion, so that nothing else hath part therein: and the Board of the Adversaries swimmeth, when it is conjoyned with as much Air, as, together with it, doth form a Body less grave than so much water as would fill the place that the said Compound occupyes in the water; but when they shall demit the simple Ebony into the water, according to the Tenour of our Question, it shall alwayes go to the bottom, though it were as thin as a Paper.

---

[12] Lib. 4. c. 6. Text 45.

FINIS.

.

Also from Benediction Books ...
Wandering Between Two Worlds: Essays on Faith and Art
Anita Mathias
Benediction Books, 2007
152 pages
ISBN: 0955373700

Available from www.amazon.com, www.amazon.co.uk

In these wide-ranging lyrical essays, Anita Mathias writes, in lush, lovely prose, of her naughty Catholic childhood in Jamshedpur, India; her large, eccentric family in Mangalore, a sea-coast town converted by the Portuguese in the sixteenth century; her rebellion and atheism as a teenager in her Himalayan boarding school, run by German missionary nuns, St. Mary's Convent, Nainital; and her abrupt religious conversion after which she entered Mother Teresa's convent in Calcutta as a novice. Later rich, elegant essays explore the dualities of her life as a writer, mother, and Christian in the United States-- Domesticity and Art, Writing and Prayer, and the experience of being "an alien and stranger" as an immigrant in America, sensing the need for roots.

About the Author

Anita Mathias is the author of *Wandering Between Two Worlds: Essays on Faith and Art.* She has a B.A. and M.A. in English from Somerville College, Oxford University, and an M.A. in Creative Writing from the Ohio State University, USA. Anita won a National Endowment of the Arts fellowship in Creative Nonfiction in 1997. She lives in Oxford, England with her husband, Roy, and her daughters, Zoe and Irene.

Visit Anita's website
http://www.anitamathias.com,
and Anita's blog
http://dreamingbeneaththespires.blogspot.com, (Dreaming Beneath the Spires).

The Church That Had Too Much
Anita Mathias
Benediction Books, 2010
52 pages
ISBN: 9781849026567

Available from www.amazon.com, www.amazon.co.uk

The Church That Had Too Much was very well-intentioned. She
wanted to love God, she wanted to love people, but she was both
hampered by her muchness and the abundance of her posses-
sions, and beset by ambition, power struggles and snobbery.
Read about the surprising way The Church That Had Too Much
began to resolve her problems in this deceptively simple and en-
chanting fable.

About the Author

Anita Mathias is the author of *Wandering Between Two Worlds:
Essays on Faith and Art.* She has a B.A. and M.A. in English
from Somerville College, Oxford University, and an M.A. in
Creative Writing from the Ohio State University, USA. Anita
won a National Endowment of the Arts fellowship in Creative
Nonfiction in 1997. She lives in Oxford, England with her husband, Roy, and her daughters, Zoe and Irene.

Visit Anita's website
   http://www.anitamathias.com,
and Anita's blog
   http://dreamingbeneaththespires.blogspot.com (Dreaming Beneath the Spires).